A-Z
OPTICS

A-Z OPTICS

Prof. G.K. Bose

CENTRUM PRESS
NEW DELHI-110002 (INDIA)

CENTRUM PRESS
H.O.: 4360/4, Ansari Road, Daryaganj,
New Delhi-110 002 (India)
Ph.: 23278000, 23261597

B.O.: No. 1015, Ist Main Road, BSK IIIrd Stage
IIIrd Phase, IIIrd Block,
Bangalore - 560 085 (India)
Tel.: 080-41723429
Visit us at: www.centrumpress.com

A-Z Optics

First Edition, 2009

ISBN 978-93-80106-30-4

PRINTED IN INDIA

Printed at Salasar Imaging Systems, Delhi-110035 (India)

Contents

Preface

Providing an accurate quantitative description of wave propagation through a complex medium remains one of the more difficult mathematical problems of our age. In electrodynamics, many alternative-solution techniques for Maxwell's equations have emerged over the past century; all involve approximations in some form or another. Moreover, particularly when the wavelength of the wave is small compared to the scale of the refracting medium, one is forced to analytic and geometric simplifications to render such problems tractable.

The radio occultation technique using spacecraft to sound a refracting medium is a relative newcomer in wave propagation processes and in the inverse problem of inferring certain physical properties of the medium. The Mariner radio signal received on the Earth first transected the Martian ionosphere and atmosphere during its immersion before being eclipsed by Mars, and it passed through these media again after its egress from behind Mars.

Using ray theory techniques, partly borrowed from seismology, these initial experiments successfully recovered accurate vertical refractivity profiles for the Martian atmosphere, and related density and pressure information. As anyone who seriously studies sunrises and sunsets could predict, dense atmospheres with even simple mesocale layered structures, not to mention more complex structures, can lead to difficulty in the inversion process using basic ray theory techniques.

Author

Chapter 1

Light

Light may be defined subjectively as the sense-impression formed by the eye. This is the most familiar connotation of the term, and suffices for the discussion of optical subjects which do not require an objective definition, and, in particular, for the treatment of physiological optics and vision. The objective definition, or the "nature of light," is the *ultima Thule* of optical research. "Emission theories," based on the supposition that light was a stream of corpuscles, were at first accepted.

These gave place during the opening decades of the 19th century to the "undulatory or wave theory," which may be regarded as culminating in the "elastic solid theory "- so named from the lines along which the mathematical investigation proceeded - and according to which light is a transverse vibratory motion propagated longitudinally though the aether.

The mathematical researches of James Clerk Maxwell have led to the rejection of this theory, and it is now held that light is identical with electromagnetic disturbances, such as are generated by oscillating electric currents or moving magnets. Beyond this point we cannot go at present. To quote Arthur Schuster, "So long as the character of the displacements which constitute the waves remains undefined we cannot pretend to have established a theory of light."

Optical and electrical phenomena are co-ordinated as a phase of the physics of the "aether," and that the investigation of these sciences culminates in the derivation of the properties of this conceptual medium, the existence of which was called into being as an instrument of research.' The methods of the elastic-solid theory can still be used with advantage in treating many optical phenomena, more especially so long as we remain ignorant of fundamental matters concerning the origin of electric and magnetic strains and stresses; in addition, the treatment is more intelligible, the researches on the electromagnetic theory leading in many cases to the derivation of differential equations which express quantitative relations between diverse phenomena, although no precise meaning can be attached to the symbols employed.

It is the fastest in the universe, and has to be the fastest, since nothing can even reach its speed. It can split into colours, and can be recombined from colours. In fact, a universe without it cannot even be imagined.Light, or "Adustum", as it is known in Latin,is a most apt title for our web site. Light is the ultimate form of energy. It is not just about the visible wavelengths, but even more about the numerous others, that sustain life on earth.

White light received from the Sun has all these wavelengths, and forms a much wider electromagnetic spectrum, that the visible spectrum. Well, that is what our site is all about - not just visible light, but light as a whole.Each of the six main links in the site come with a set of sub-links that deal with individual topics. Besides text, there is a huge collection of diagrams and images that have been prepared and collected to make the matter more lucid and interesting.

Now what exactly is light? A Scottish physicist named James Clerk Maxwell showed that electric and magnetic fields fluctuating together can form a propagating wave, which was named an electromagnetic wave. Light is this type of wave. Maxwell knew that a changing electric field produced a magnetic field. He hypothesized that a changing magnetic field would produce an electric field because most things in nature are balanced. Based on the work of Farady and Hertz, this was discovered this to be true.

A common way to generate an electromagnetic wave is to have an antenna connected to an alternating current which will cause a changing magnetic field which will cause a changing electric field which will cause another changing magnetic field and so on. These waves are also produced in stars in the form of UV rays, X rays and Gamma rays.

An electromagnetic wave, like other waves, has a frequency f, a speed v and a wavelength λ, which are related by the equation $v = f\lambda$. In vacuum or in air, to a good approximation, the speed $v = c$ (speed of light: 3×10^8 m/s) so the relationship would be $c = f\lambda$

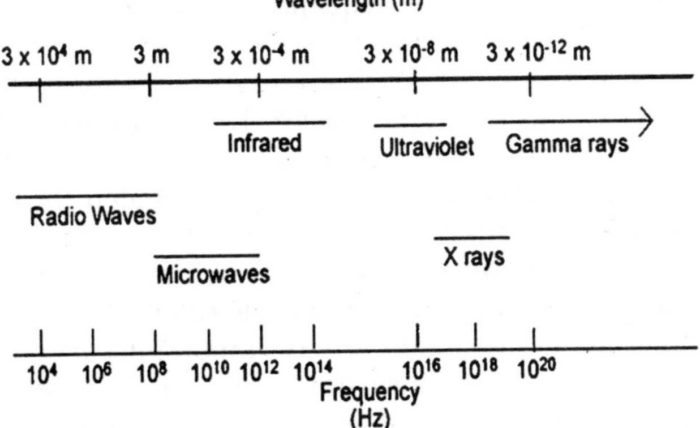

Above shows the electromagnetic spectrum, which depicts electromagnetic waves over a huge range, for values less than 104 Hz to those greater than 1022 Hz. Because all of these waves travel at the same speed, this drawing can also be used to figure out the wavelengths. The grouping of the spectrum into radio waves, infrared, ultraviolet, x and gamma rays. Shown here the boundaries are sharp but in actual practice the boundaries are not so well defined and the regions overlap.

Beginning on the left are the radio waves. The lower frequency radio waves are generally produced by electric oscillator circuits while the higher frequency waves (microwaves) are generated using electron tubes called klystrons.

The next set, called infrared or loosely referred to as heat waves, generally originate with the vibration and rotation of molecules within a material. The next set,which we are most familiar with, is the visible light waves, represented by a white stripe. These are generally emitted by hot objects, such as the sun or the filament of an incandescent light bulb where the temperature is high enough to excite the electrons within an atom. Ultraviolet rays are generally produced from the discharge of an electric arc and x-rays are generated by the sudden deceleration of high-speed electrons. And last, gamma rays are radiation from nuclear decay.

Of all the frequencies, the most familiar set, that of light, is the smallest range indicated in the spectrum. The human eye can perceive only the frequencies between 4.0×10^{14} Hz and 7.9×10^{14} Hz as visible light. For the most part, visible light is discussed in terms of wavelengths with the unit being nanometers (nm) where 1 nm = 10^{-9} m rather than in frequencies. Our brain recognizes these wavelengths as different colors where the 750 nm is approximately the longest wavelength of red light and 380 nm is approximately the shortest wavelength of violet light. In between these two wavelengths are all of the other visible colors.

Wavelength also plays a very important part on other parts of the magnetic spectrum besides defining color for visible light. One important aspect about wavelength is that it determines how much a wave diffracts, or bends around an obstacle. Longer wavelengths bend around an obstacle more than shorter wavelengths. An example of why this is important is with radio waves. AM waves are significantly longer than FM waves and therefore they have a greater ability to bend around buildings than FM waves do.The inability of FM waves to diffract around obstacles is one of the major reasons why FM stations broadcast their signals from high areas in a "line-of-sight" manner.

HISTORY OF LIGHT

Many of the early scientists who studied sound also studied light. Pythagoras, for example, believed that light came from visible objects toward the eye. However, in addition to this basically correct thought, Plato and many other Greeks, also held the mistaken belief that vision issued out from the eye. Despite this, many of the ideas of the ancient Greeks were accurate. The philosopher Empedocles correctly believed that light traveled with finite speed. Aristotle, too, conjectured about light as well as sound. He rightly explained rainbows as a sort of reflection off of raindrops. The mathematician Euclid worked with mirrors and reflection, and many other thinkers observed refraction, though they did not know how to express it mathematically.

Ptolemy is the first recorded person to experiment with optics and collect data, but he believed in Plato's mistaken thought that vision extended out from the eye. Ptolemy's work was further developed by the Egyptian scientist Ibn al Haythen, who was known to Europeans as Alhazen. It was Alhazen who first drew ray diagrams, and who managed to discount Plato's theory, through a mixture of logic and experimentation. His work, done during the time of the Moorish Empire, was influential in later studies of light.

Many advances in the study of light were made in the 16th and 17th centuries by such renowned scientists as Galileo Galilei, Johannes Kepler, and Renes Descartes. Both Descartes and Dutch mathematician Willebrord Snell independently developed the law of refraction, basing much of their work on the earlier work of Alhazen. Snell discovered it earlier, however, and the law is now known as Snell's law.

During the late 17th century, a debate grew over whether light behaved as a particle or a wave. Sir Isaac Newton believed in a particle theory of light, in part because he did not observe diffraction in light, a property it should have had, had it been a wave. Though he did not know it, this was due a lack of coherency in the light when he was experimenting. Coherency is important, because without a steady, coherent beam, it is impossible to observe such things as diffraction of light. These days, coherency is generally achieved with a laser, a modern invention that was unavailable to Newton. There were those among his contemporaries, however, who believed in a wave theory of light. The most notable among these was the Dutch scientist Christiaan Huygens, who first wrote of light as a wave.

Though Newton had difficulty explaining certain phenomena of light, it seemed that the wave theory also had difficulty explaining certain optical phenomena. Because of Newton's prestige, the particle theory was accepted for almost a century.

It was not until the early 19th century that the wave theory of light became widely accepted. This acceptance came in large part due to the work of the English doctor Thomas Young. In 1801 he reported his famous double-slit experiment, which clearly showed light to diffract, and thus travel as a wave. By passing light through a single slit before passing it through a double slit, he managed to emulate a point source of light, and to achieve the coherency that had eluded Newton. In the 1850s Fizeau and Foucault showed through measurements that light traveled slower through denser media. In the same century, Augustin Fresnel and later James Clerk Maxwell, working within a wave theory of light seemed to explain phenomena that Newton had been unable to explain in terms of a particle theory of light, such as polarization, interference, and diffraction.

They were also able to address the question of what determines which part of the light will be reflected and which transmitted when light is reflected at a surface such as glass or water. This question was opened back up in the 20th century with the rise of quantum theory.

It was the Irish mathematician Sir William Hamilton who developed a theory that joined optics and mechanics, thus elucidating the relationship between wave and particle viewpoints. His theory helped lay the foundation for the later development of quantum mechanics

James Clerk Maxwell did much more to aid the study of light. In essence, he was the father of our modern perception of light.

Working from Michael Faraday's discoveries in electricity and magnetism, he managed to fully explain electromagnetic phenomena, developing a theory that described how electromagnetic waves traveled through space. His theory was later experimentally proven correct by Heinrich Rudolph Hertz, the man whose name gives us the unit hertz, one of which is equivalent to one cycle per second.

The velocity of the electromagnetic waves that Maxwell had described was found to be the same as that of light, essentially proving that light was an electromagnetic wave.Maxwell's work seemed to have finally settled the issue of whether light was a wave or a particle, but the whole debate was reopened in the 20th century.

Scientists such as Albert Einstein, who described the Doppler effect for light, brought the particle theory back into the picture with quantum theory. This time they postulated that light did not just behave as a particle or a wave, but had properties of both.

RAY OPTICS

What are the physics of light? The best way to describe how light travels would be to use *ray optics*. Rays are basically infinitesimally thin slices of waves in the direction that they are traveling.

Reflection

Practically all objects reflect a certain portion of the light falling on them. The light that is reflected off an object is the color that the eyes see. If a ray of light hits a flat, shiny surface, than it reflects off at the same angle. This means the ray that hits the surface, called the *incident ray,* which has an angle of incidence θ_i from the *normal* (an imaginary line perpendicular to the surface at the point where the incident ray encounters it) bounces off as the *reflected ray,* which has an equal angle of reflection θ_r from the normal.

This is leads to the law of reflection which follows:

$$\theta_r = \theta_i$$

The incident ray, the reflected ray, and the normal to the surface all lie in the same plan, and the angle of reflection, , equals the angle of incidence, θ_i. There are two types of reflection. *Specular reflection* is when parallel light rays strike a smooth, plane surface, and reflect off parallel.

This is important in determining the properties of mirrors because it shows how much the mirror reflects without distortion.

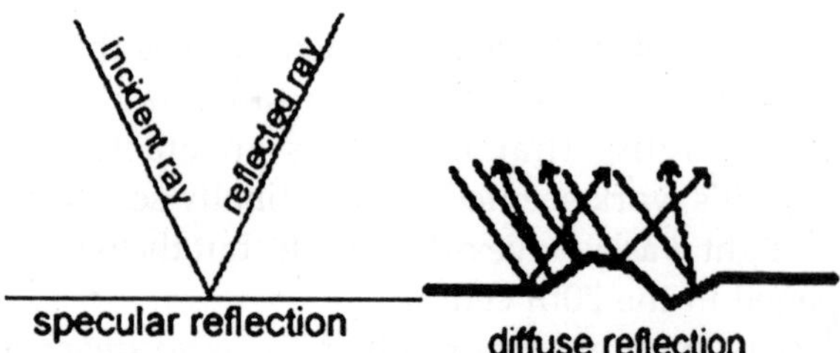

The other type of reflection, *diffuse reflection,* is when parallel rays strike a rough surface, causing them to reflect in all directions, causing distortion.

Images Formed by a Plane Mirror

Looking into a plane mirror, you can see an image of yourself that has four fundamental properties:

- The image is left to upright.
- The image is the same size as you are.
- The image is located as far behind the mirror as you are in front of it
- The image is reversed left to right.

The image being reversed means that when you wave your right hand, the image seems to wave its left hand. The image you see that is behind the mirror is called the *virtual image.*

Since the light rays do not really pass through the image, images formed by a plane mirror are known as virtual images. You can distinguish them from *real images,* which are formed by light that does pass through the image because a white piece of paper placed at the image location will not show anything if the image is virtual, but will show the image if the image is real.

Parabolic Mirrors

A parabolic mirror is a common type of curved mirror. In general shape, the mirror is roughly a section of the surface of a sphere, though as the name suggests, the curve is not perfectly spherical. It is parabolic. There are two types of parabolic mirrors: concave and convex. Imagine a parabola with the inside surface (the part on the inside of the curve) as a reflective surface. This is a concave parabolic mirror. Conversely, a parabola with the outside as a reflective surface is a convex parabolic mirror.

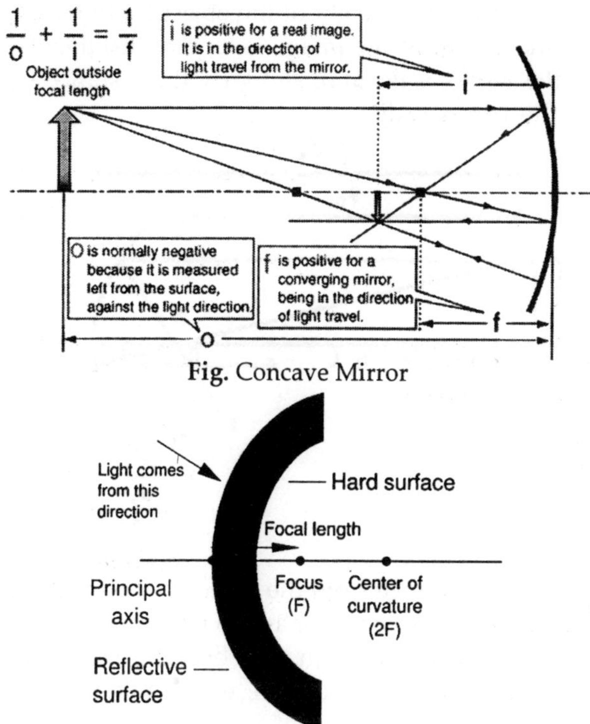

Fig. Concave Mirror

Fig. Convex Mirror

The law of reflection applies two both these types of mirrors as it did with the plane mirror. Only now the normal is drawn perpendicular to the mirror at the point of incidence, which is where the ray strikes the surface of the mirror. Both these types of mirrors have a vertex, shown on the diagram. The line perpendicular to the surface of the mirror at the vertex is called the principal axis.

The ray parallel to the principle axis goes through a point labeled F.

This is the focal point of the mirror. All rays of light parallel to the principle axis reflecting off the surface of the mirror will go through this point. Similarly, a ray going through the focal point will be reflected back parallel to the principal axis.

The focal length of the mirror, *f*, is the distance from the focus to the vertex. The height of the object is otherwise stated *ho*, while the height of the image is *hi*. The distance from the object to the vertex is *do*, and, similarly the distance from the image to the vertex is *di*.

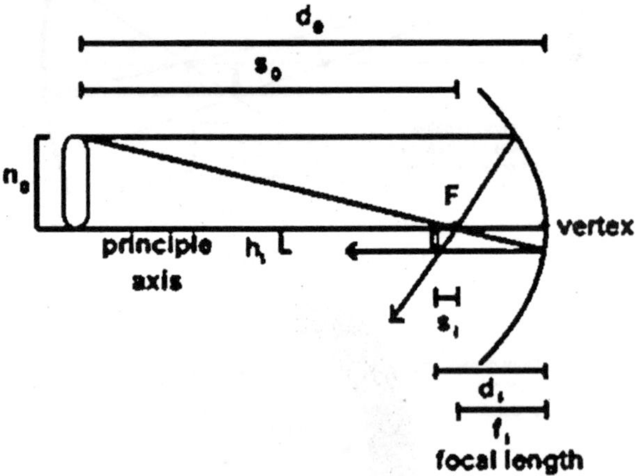

Finally, the distance from the object to the focal point F is *so*, and the distance from the image to the focus is *si*. We label all these distances for ease in use, because there are many useful relationships between them, which follow.

$$\frac{ho}{hi} = \frac{so}{f} = \frac{do}{di} = \frac{f}{si} \text{ as well as } \frac{1}{f} = \frac{1}{di} + \frac{1}{do}.$$

Refraction

Light will always travel in the fastest path from point A to point B. If this path goes through different materials through which light travels at different speeds, then light will not travel in a straight line.

This bending of light when it enters a new medium is known as refraction, and is the reason a stick looks bent when we hold part of it in water. Our eyes are trained to think that light travels to us in a straight line, so when it bends, we think it is the object that is bent, not the light.

Light travels faster in media that are less optically dense and slower in media that are more optically dense. The index of refraction is the ratio of the speed of light in a vacuum to the speed of light in the medium, and is represented by the letter n.

$$\cdot\, n = \frac{\text{Speed of light in a vaccum}}{\text{Speed of light in the medium}} = \frac{c}{v}$$

This index of refraction is quite important, because we can use it to determine at what angle the light bends. We do this using an equation developed by Wilebrord Snell, named, aptly enough, Snell's law. When light hits the boundary between two transparent media, it generally splits into two parts. One part reflects back into the first medium, the other refracts into the second media. The exception to this is when the light hits at such an angle that there is total internal reflection, which means that all the light bounces back into the original medium.

Snell's law of refraction states that when light travels from a medium with a refractive index of n_1 into a medium with a refractive index of n_2, the incident ray, normal ray, and reflected ray all lie in the same plane. Furthermore, θ, (the angle of incidence) and (the angle of refraction are related by the following formula:

$$n_1 \sin = \sin \theta_1 = n_2 \sin \theta_2$$

The speed of light in air is so close to the speed of light in a vacuum, that for most purposes we can use a value of one for the index of refraction of air. Light travels more noticeabley slowly in some other materials, however, and a table of some of these values follows. Of course, light can never travel faster than c, its speed in a vacuum, which is m/s.

Substance (at 20° C)	*Index of refraction, n*
Diamond	2.419
Sodium Chloride	1.544
Water	1.333
Benzene	1.501

EXPLANATION OF REFLECTION AND REFRACTION

Newton supposed the light-corpuscles to be subjected to attractive and repulsive forces exerted at very small distances by the particles of matter. In the interior of a homogeneous body a corpuscle moves in a straight line as it is equally acted on from all sides, but it changes its course at the boundary of two bodies, because, in a thin layer near the surface there is a resultant force in the direction of the normal. In modern language we may say that a corpuscle has at every point a definite potential energy, the value of which is constant throughout the interior of a homogeneous body, and is even equal in all bodies of the same kind, but changes from one substance to another.

If, originally, while moving in air, the corpuscles had a definite velocity vo, their velocity v in the interior of any other substance is quite determinate. It is given by the equation 2mv 2 -1mv 0 2 =A, in which m denotes the mass of a corpuscle, and A the excess of its potential energy in air over that in the substance considered.

A ray of light falling on the surface of separation of two bodies is reflected according to the well-known simple law, if the corpuscles are acted on by a sufficiently large force directed towards the first medium. On the contrary, whenever the field of force near the surface is such that the corpuscles can penetrate into the interior of the second body, the ray is refracted. In this case the law of Snellius can be deduced from the consideration that the projection w of the velocity on the surface of separation is not altered, either in direction or in magnitude. This obviously requires that the plane passing through the incident and the refracted rays be normal to the surface, and that, if a, and α^2 are the angles of incidence and of refraction, v, and v_2 the velocities of light in the two media, $\sin \alpha^1 / \sin \alpha^2 = w/v : w/v^2 = v^2/v$. The ratio is constant, because, as has already been observed, v_1 and v_2 have definite values.

As to the unequal refrangibility of differently coloured light, Newton accounted for it by imagining different kinds of corpuscles. He further carefully examined the phenomenon of total reflection, and described an interesting experiment connected with it. A marked change is observed when a second piece of glass is made to approach the reflecting face, so as to be separated from it only by a very thin layer of air.

The reflection is then found no longer to be total, part of the light finding its way into the second piece of glass. Newton concluded from this that the corpuscles are attracted by the glass even at a certain small measurable distance.

New Hypotheses in the Corpuscular Theory

The preceding explanation of reflection and refraction is open to a very serious objection. If the particles in a beam of light all moved with the same velocity and were acted on by the same forces, they all ought to follow exactly the same path. In order to understand that part of the incident light is reflected and part of it transmitted, Newton imagined that each corpuscle undergoes certain alternating changes; he assumed that in some of its different "phases" it is more apt to be reflected, and in others more apt to be transmitted. The same idea was applied by him to the phenomena presented by very thin layers.

He had observed that a gradual increase of the thickness of a layer produces periodic changes in the intensity of the reflected light, and he very ingeniously explained these by his theory. It is clear that the intensity of the transmitted light will be a minimum if the corpuscles that have traversed the front surface of the layer, having reached that surface while in their phase of easy transmission, have passed to the opposite phase the moment they arrive at the back surface.

As to the nature of the alternating phases, Newton expresses himself as follows: "Nothing more is requisite for putting the Rays of Light into Fits of easy Reflexion and easy Transmission than that they be small Bodies which by their attractive Powers, or some other Force, stir up Vibrations in what they act upon, which Vibrations being swifter than the Rays, overtake them successively, and agitate them so as by turns to increase and decrease their Velocities, and thereby put them into those Fits."

The Corpuscular Theory

Though Newton introduced the notion of periodic changes, which was to play so prominent a part in the later development of the wave-theory, he rejected this theory in the form in which it had been set forth shortly before by Christiaan Huygens in chief objections being:

- That the rectilinear propagation had not been satisfactorily accounted for;
- That the motions of heavenly bodies show no sign of a resistance due to a medium filling all space; and
- That Huygens had not sufficiently explained the peculiar properties of the rays produced by the double refraction in Iceland spar. In Newton's days these objections were of much weight.

Yet his own theory had many weaknesses. It explained the propagation in straight lines, but it could assign no cause for the equality of the speed of propagation of all rays. It adapted itself to a large variety of phenomena, even to that of double refraction but it could do so only at the price of losing much of its original simplicity. In the earlier part of the 19th century, the corpuscular theory broke down under the weight of experimental evidence, and it received the final blow. Foucault proved by direct experiment that the velocity of light in water is not greater than that in air, as it should be according to the formula but less than it, as is required by the wave-theory.

Light as a Wave

The classical description of light as an electromagnetic wave makes some assumptions about its nature, based on what could be observed at the time. Although some of these assumptions have proven to be less than accurate as we learn more about electromagnetic phenomena, they make a good starting point for our discussion about light in particular and electromagnetic waves in general. The two basic assumptions were:

The frequency of the electromagnetic wave can be varied over the entire positive range, but cannot be reduced to zero.

This seems intuitively obvious. A frequency of zero would mean no change in the strengths of the electric and magnetic fields. But an electromagnetic wave reqires these fields to be constantly changing in order to exist. And the idea of a negative frequency seems ludicrous.

The energy in the wave is continuously variable, with a minimum energy of any non-zero value, and no maximum value.

This also seems intuitive. Looking at sunlight and comparing light intensity on a clear day and a cloudy day, we note that the clouds block some of the sun's energy. The desert sun at Noon is very intense, while the setting sun at extreme northern or southern latitudes is much less noticeable. Yet it is the same sunlight, coming from the same source, so we know that various factors encountered by sunlight must be removing some of the energy from it feel, and scientifically measure the difference.

Of course, other properties of electromagnetic radiation have also been determined, and a number of theories and assumptions have been developed. These have been either confirmed or disproven by experiment. Before we look at the circumstances under which the wave model of light (or any electromagnetic wave) may break down, however, let's look at the wave model itself.Since light was first recognized scientifically as a manifestation of electromagnetic energy, it can be represented as a waveform, like this:

The electrical energy present in the light waveform as it travels in the direction of the arrow, it looks as if the energy level is becoming alternately positive and negative, with momentary crossovers of zero electrical energy. In the basic model of light shown here, this is in fact the case; as with all electromagnetic waves, light energy is constantly changing its form between electrical energy and magnetic energy. The point of maximum magnetic energy coincides with the moment of zero electrical energy.

Beyond that instant, energy shifts again from the magnetic field back into the electrical field, but with a reversal from the previous polarity. This continues as long as that particular ray of light exists.There are other "modes" of propogation which involve more complex interactions between the electric and magnetic fields, but in all cases the Law of Conservation necessarily holds true:

Energy is neither created nor destroyed as it is transformed from one form to another; the total energy in the wave must somehow remain constant throughout the full cycle.

The sine wave shown here represents the strength and polarity of the electrical field associated with the motion of this ray of light. The light itself, assuming no outside influences, travels in the straight line indicated by the blue arrow. The light energy does *not* "wiggle" back and forth as it moves along its path.As an electromagnetic wave, light has some characteristics in common with all forms of electromagnetic energy. These include *wavelength, frequency,* and *speed of propogation*. These characteristics are actually related to each other, so that any one can be calculated if the other two are known. Let's take a look at each of these characteristics:

Wavelength

Since light is a repeating waveform in motion, it is possible to measure the physical distance between matching points of adjacent cycles of the waveform. This is shown here:

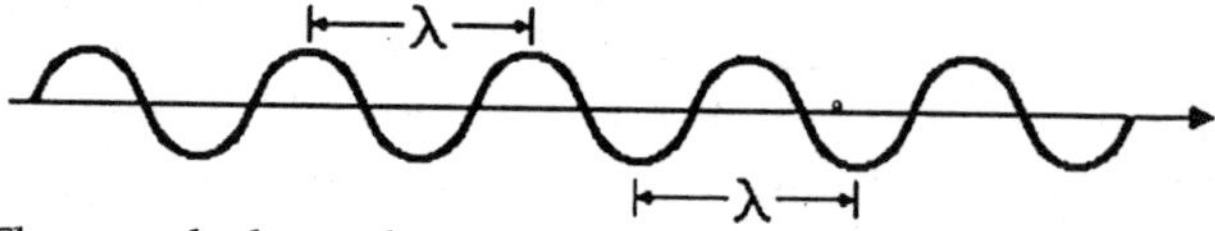

The symbol used to represent this distance is the Greek letter "Lambda" (λ).

The wavelength can actually be measured between any two corresponding points on the waveform. It is convenient to use the most positive point or the most negative point, both of which are shown above. However, we could have just as easily specified two zero-crossing points, so long as both crossed the zero line in the same direction.Remember that the light itself does not wave back and forth along its path of travel.

What we are actually measuring here is the distance traveled through space by this ray of light, while its electrical field goes from its maximum positive value, through zero to its maximum negative value, and then through zero again to once more reach its maximum positive value.This distance is normally measured in meters (m) or some decimal fraction of a meter, such as centimeters (cm). The correct units of measurement are meters per cycle (m/cycle) or some appropriate derivation. In the case of light, the wavelength is so short that a specific distance, called the *angstrom* (Å), has been defined.

One ångstrom = 10^{-10} m or 10^{-8} cm.

Visible light has a characteristic wavelength in the range of approximately 3900 Å to 7700 Å. Electromagnetic energy outside this range is no longer visible to the human eye.

Speed of Propogation

The speed at which light travels through any medium is determined by the density of that medium. The presence of matter, even transparent matter, will slow the light down. Even air will have some effect, and glass has a more significant effect on the speed at which light will travel through it.Ever more sophisticated experiments have determined the speed of light quite accurately. According to current knowledge:

Speed of light in a vacuum = $2.997925 \pm 0.000002 \times 10^{10}$ cm/sec.

As made famous in Einstein's equation, the letter c is used as a general symbol for the speed of light.

Frequency and Period

In any electromagnetic wave, it takes time for the energy in the wave to change from electrical format to magnetic and then back again. The amount of time required to do this twice, covering one complete cycle, or wavelength of the signal is known as the *period* of the wave. Thus, the period of any wave, measured as some amount of time per cycle, is in fact the time interval that corresponds to the physical wavelength of the signal.

The *frequency* of the wave is the inverse or reciprocal of the period. That is, the frequency is the number of cycles of the waveform that occur in one second of time. For many years this was simply measured in units of cycles per second. Recently, however, the specific name *hertz* (abbreviated Hz) has been designated as the appropriate unit to indicate cycles per second.In general equations, the letter f is used to indicate frequency in hertz.The basic mathmatical formula that relates wavelength, frequency, and the speed of light is:

$$c = f\lambda$$

The wave theory of light was happily adopted and accepted until it was found to fail to explain some observed and measured phenomena, in consistent and repeatable experiments. The two phenomena that upset this model are the photoelectric effect and blackbody radiation. These two effects could only be explained by assuming that light energy propogates as a series of independent "corpuscles," or bundles. This gave rise to the more recent particle theory of light.

GENERAL THEOREMS ON RAYS

- *Let A and B be two points arbitrarily chosen in a system of transparent bodies,* ds *an element of a line drawn from A to B, u the velocity of a ray of light coinciding with* ds. *Then the integral* fulds, *which represents the time required for a motion along the line with the velocity* u, *is a minimum for the course actually taken by a ray of light (unless A and B be too far apart). This is the " principle of least time " first formulated by Pierre de Fermat for the case of two isotropic substances. It shows that the course of a ray of light can always be inverted.*
- Rays of light starting in all directions from a point A and travelling onward for a definite length of time, reach a surface *a,* whose tangent plane at a point B is conjugate, in the medium surrounding B, with the last element of the ray AB.
- If all rays issuing from A are concentrated at a point B, the integral *fu 'ds* has the same value for each of them.
- In case the variation of the integral caused by an infinitely small displacement *q* of B, the point A

remaining fixed, is given by *Sfu l ds=q* cos *0/v B* . Here 0 is the angle between the displacement *q* and the normal to the surface *a,* in the direction of propagation, vs the velocity of a plane wave tangent to this surface.

Further General Theorems

- Let V_1 and be two planes in a system of isotropic bodies, let rectangular axes of coordinates be chosen in each of these planes, and let x_i, y_1 be the coordinates of a point A in V_1, and those of a point B in The integral *fads,* taken for the ray between A and B, is a function of x_1, and, if 51 denotes either x_1 or y_1, and 12 either x_2 or y_2 we shall have as = α_1. On both sides of this equation the first differentiation may be performed by means of the formula. The second differentiation admits of a geometrical interpretation, and the formula may finally be employed for proving the following theorem: Let ω_1 be the solid angle of an infinitely thin pencil of rays issuing from A and intersecting the plane in an element α^2 at the point B. Similarly, let ω_2 be the solid angle of a pencil starting from B and falling on the element *a l* of the plane V_1 at the point A. Then, denoting by 111 and 112 the indices of refraction of the matter at the points A and B, by 01 and 02 the sharp angles which the ray AB at its extremities makes with the normals to V_1 and V_2, we have 1 1 1 α^1 = 1122 α^2 ω_2 cos
- There is a second theorem that is expressed by exactly the same formula, if we understand by α^1 and α^2 elements of surface that are related to each other as an object and its optical image - by col, ω^2 the infinitely small openings, at the beginning and the end of its course, of a pencil of rays issuing from a point A of *ai* and coming together at the corresponding point B of α^2, and by 0 1, the sharp angles which one of the rays makes with the normals

to *al* and α^2. The proof may be based upon the first theorem. It suffices to by a line drawn from the centre of the wave-surface to its point of contact with a tangent plane of the given direction. It will be convenient to say that this line and the plane are conjugate with each other. The rays of light, curved in non-homogeneous bodies, are always straight lines in homogeneous substances. In an isotropic medium, whether homogeneous or otherwise, they are normal to the wave-fronts, and their velocity is equal to that of the waves.

By applying his construction to the reflection and refraction of light, Huygens accounted for these phenomena in isotropic bodies as well as in Iceland spar. rt was afterwards shown by Augustin Fresnel that the double refraction in biaxal crystals can be explained in the same way, provided the proper form be assigned to the wavesurface.

In any point of a bounding surface the normals to the reflected and refracted waves, whatever be their number,. always lie in the plane passing through the normal to the incident waves and that to the surface itself. Moreover, if a l is the angle between these two latter normals, and α^2 the angle between the normal to the boundary and that to any one of the reflected and refracted waves, and v, v_2 the corresponding wave-velocities, the relation sin a l /sin $\alpha^2 = v_1/v_2$ is found to hold in all cases.

These important theorems may be proved independently of Huygens's construction by simply observing that, at each point of the surface of separation, there must be a certain connexion between the disturbances existing in the incident, the reflected, and the refracted waves, and that, therefore, the lines of intersection of the surface with the positions of an incident wavefront, succeeding each other at equal intervals of time *dt*, must coincide with the lines in which the surface is intersected by a similar series of reflected or refracted wave-fronts.

In the case of isotropic media, the ratio is constant, so that we are led to the law of Snellius, the index of refraction being given by consider the section a of the pencil by some intermediate plane, and a bundle of rays starting from the points of a l and reaching those of *a 2* after having all passed through a point of that section *o*-. If in the last theorem the system of bodies is symmetrical around the straight line AB, we can take for a l and $6r_2$ circular planes having AB as axis. Let *h l* and h_2 be the radii of these circles, *i.e.* the linear dimensions of an object and its image, e_i and e_2 the infinitely small angles which a ray R going from A to B makes with the axis at these points.

Then the above formula gives / 1, h_1 e_i =A2$h_2 e_2$, a relation that was proved, for the particular case =P2 by Huygens and Lagrange. It is still more valuable if one distinguishes by the algebraic sign of h_2 whether the image is direct or inverted, and by that of e_2 whether the ray R on leaving A and on reaching B lies on opposite sides of the axis or on the same side. The above theorems are of much service in the theory of optical instruments and in the general theory of radiation.

Phenomena of Interference and Diffraction.

The impulses or motions which a luminous body sends forth through the universal medium or aether, were considered by Huygens as being without any regular succession; he neither speaks of vibrations, nor of the physical cause of the colours. The idea that monochromatic light consists of a succession of simple harmonic vibrations like those represented by the equation, and that the sensation of colour depends on the frequency, is due to Thomas Young' and Fresnel, 2 who explained the phenomena of interference on this assumption combined with the principle of super-position.

In doing so they were also enabled to determine the wave-length, ranging from o 000076 cm. at the red end of the spectrum to o 000039 cm. for the extreme violet and, by means of the formula, the number of vibrations per second. Later investigations have shown that the infra-red rays as well as the ultra-violet ones are of the same physical nature as the luminous rays, differing from these only by the greater or smaller length of their waves. The wave-length amounts to 0 o06 cm. for the least refrangible infra-red, and is as small as o 00001 cm. for the extreme ultraviolet.

Another important part of Fresnel's work is his treatment of diffraction on the basis of Huygens's principle. If, for example, light falls on a screen with a narrow slit, each point of the slit is regarded as a new centre of vibration, and the intensity at any point behind the screen is found by compounding with each other the disturbances coming from all these points, due account being taken of the phases with which they come together.

Results of Later Mathematical Theory

Though the theory of diffraction developed by Fresnel, and by other physicists who worked on the same lines, shows a most beautiful agreement with observed facts, yet its foundation, Huygens's principle, cannot, in its original elementary form, be deemed quite satisfactory. The general validity of the results has, however, been confirmed by the researches of those mathematicians who investigated the propagation of vibrations in a more rigorous manner. Kirchhoff 3 showed that the disturbance at any point of the aether inside a closed surface which contains no ponderable matter can be represented as made up of a large number of parts, each of which depends upon the state of things at one point of the surface.

This result, the modern form of Huygens's principle, can be extended to a system of bodies of any kind, the only restriction being that the source of light be not surrounded by the surface. Certain causes capable of producing vibrations can be imagined to be distributed all over this latter, in such a way that the disturbances to which they give rise in the enclosed space are exactly those which are brought about by the real source of light.

RAYS OF LIGHT

In working out the theory of diffraction it is possible to state exactly in what sense light may be said to travel in straight lines. Behind an opening *whose width is very large in comparison with the wave-length* the limits between the illuminated and the dark parts of space are approximately determined by rays passing along the borders.

This conclusion can also be arrived at by a mode of reasoning that is independent of the theory of diffraction.' If linear differential equations admit a solution of the form with A constant, they can also be satisfied by making A a function of the coordinates, such that, in a wave-front, it changes very little over a distance equal to the wave-length X, and that it is constant along each line conjugate with the wave-fronts. In cases of this kind the disturbance may truly be said to travel along lines of the said direction, and an observer who is unable to discern lengths of the order of X, and who uses an opening of much larger dimensions, may very well have the impression of a cylindrical beam with a sharp boundary.

A similar result is found for curved waves. If the additional restriction is made that their radii of curvature be very much larger than the wave-length, Huygens's construction may confidently be employed. The amplitudes all along a ray are determined by, and proportional to, the amplitude at one of its points.

POLARIZED LIGHT

As the theorems used in the explanation of interference and diffraction are true for all kinds of vibratory motions, these phenomena can give us no clue to the special kind of vibrations in light-waves. Further information, however, may be drawn from experiments on plane polarized light. The properties of a beam of this kind are completely known when the position of a certain plane passing through the direction of the rays, and *in* which the beam is said to be polarized, is given. " This plane of polarization," as it is called, coincides with the plane of incidence in those cases where the light has been polarized by reflection on a glass surface under an angle of incidence whose tangent is equal to the index of refraction (Brewster's law).

The researches of Fresnel and Arago left no doubt as to the direction of the vibrations in polarized light with respect to that of the rays themselves. In isotropic bodies at least, the vibrations are exactly transverse, *i.e.* perpendicular to the rays, either in the plane of polarization or at right angles to it. The first part of this statement also applies to unpolarized light, as this can always be dissolved into polarized components.

Much experimental work has been done on the production of polarized rays by double refraction and on the reflection of polarized light, either by isotropic or by anisotropic transparent bodies, the object of inquiry being in the latter case to determine the position of the plane of polarization of the reflected rays and their intensity.

In this way a large amount of evidence has been gathered by which it has been possible to test different theories concerning the nature of light and that of the medium through which it is propagated.

A common feature of nearly all these theories is that the aether is supposed to exist not only in spaces void r of matter, but also in the interior of ponderable bodies.

Fresnel's Theory

Fresnel and his immediate successors assimilated the aether to an elastic solid, so that the velocity of propagation of transverse vibrations could be determined by the formula *v=,J(K/p),* where K denotes the modulus of rigidity and p the density.

According to this equation the different properties of various isotropic transparent bodies may arise from different values of K, of p, or of both. It has, however, been found that if both K and p are supposed to change from one substance to another, it is impossible to obtain the right reflection formulae. Assuming the constancy of K Fresnel was led to equations which agreed with the observed properties of the reflected light, if he made the further assumption that the vibrations of plane polarized light are perpendicular to the plane of polarization.

As to double refraction, Fresnel made it depend on the unequal elasticity of the aether in different directions. He came to the conclusion that, for a given direction of the waves, there are two possible directions of vibration lying in the wave-front, at right angles to each other, and he determined the form of the wave-surface, both in uniaxal and in biaxal crystals. Though objections may be urged against the dynamic part of Fresnel's theory, he admirably succeeded in adapting it to the facts.

Electromagnetic Theory

James Clerk Maxwell, who had set himself the task of mathematically working out Michael Faraday's views, and who, both by doing so and by introducing many new ideas of his own, became the founder of the modern science of electricity,' recognized that, at every point of an electromagnetic field, the state of things can be defined by two vector quantities, the "electric force " E and the " magnetic force " H, the former of which is the force acting on unit of electricity and the latter that which acts on a magnetic pole of unit strength.

In a non-conductor (dielectric) the force E produces a state that may be described as a displacement of electricity from its position of equilibrium.

This state is represented by a vector D ("dielectric displacement") whose magnitude is measured by the quantity of electricity reckoned per unit area which has traversed an element of surface perpendicular to D itself. Similarly, there is a vector quantity B (the "magnetic induction") intimately connected with the magnetic force H.

Changes of the dielectric displacement constitute an electric current measured by the rate of change of D, and represented in vector notation by C==D Periodic changes of D and B may be called "electric " and "magnetic vibrations." Properly choosing the units, the axes of coordinates (in the first proposition also the positive direction of s and n), and denoting components of vectors by suitable indices, we can express in the following way the fundamental propositions of the theory.

- Let *s* be a closed line, a a surface bounded by it, n the normal to a. Then, for all bodies, *H 8 ds = f C n dcr,* J *Esds* = - *d* f *Bnda,* where the constant *c* means the ratio between the electro-magnet and the electrostatic unit of electricity. From these equations we can deduce:

For a surface of separation, the continuity of the tangential components of E and H. The solenoidal distribution of C and B, and in a dielectric that of D. A solenoidal distribution of a vector is one corresponding to that of the velocity in an incompressible fluid. It involves the continuity, at a surface, of the normal component of the vector.

- The relation between the electric force and the dielectric displacement is expressed by Dx = eiEx, Dy = e2Ey, D z = E3Ez, the constants (dielectric constants) depending on the properties of the body considered. In an isotropic medium they have a common value *e*, which is equal to unity for the free aether, so that for this medium D = E.
- There is a relation similar to between the magnetic force and the magnetic induction. For the aether, however, and for all ponderable bodies with which this article is concerned, we may write B =H.

It follows from these principles that, in an isotropic dialectric, transverse electric vibrations can be propagated with a velocity *v = c/-J E.* Indeed, all conditions are satisfied if we put Dx =o, Dy =*a* cos *n(t - xtr'+l), 'D z =o, H z =o, H y =0, Hz=* avc 1 *cos* n(t – xa+*l*) For the free aether the velocity has the value c.

Now it had been found that the ratio *c* between the two units of electricity agrees within the limits of experimental errors with the numerical value of the velocity of light in aether. (The mean result of the most exact determinations 2 of c is 3,001 to 10 cm./sec., the largest deviations being about 0,008 10 10 :. and Cornu 3 gives 3,001 to 10 o,003 t010 as the most probable value of the velocity of light.) By this Maxwell was led to suppose that light consists of transverse electromagnetic disturbances.

On this assumption, the equations represent a beam of plane polarized light. They show that, in such a beam, there are at the same time electric and magnetic vibrations, both transverse, and at right angles to each other. It must be added that the electromagnetic field is the seat of two kinds of energy distinguished by the names of electric and magnetic energy, and that, according to a beautiful theorem due to J. H. Poynting, 4 the energy may be conceived to flow in a direction perpendicular both to the electric and to the magnetic force.

The amounts per unit of volume of the electric and the magnetic energy are given by the expressions 2 (+ Ey Dy + Ez Dz), a (HxBx +Hy B y+HzBz) = whose mean values for a full period are equal in every beam of light.

The index of refraction of a body is given by result that has been verified by Ludwig Boltzmann's measurements' of the dielectric constants of gases. Thus Maxwell's theory can assign the true cause of the different optical properties of various transparent bodies.

It also leads to the reflection formulae , provided the electric vibrations of polarized light be supposed to be perpendicular to the plane of polarization, which implies that the magnetic vibrations are parallel to that plane.

Following the same assumption Maxwell deduced the laws of double refraction, which he ascribes to the unequality of E l, 21 His results agree with those of Fresnel and the theory has been confirmed by Boltzmann,' who measured the three coefficients in the case of crystallized sulphur, and compared them with the principal indices of refraction. Subsequently the problem of crystalline reflection has been completely solved and it has been shown that, in a crystal, Poynting's flow of energy has the direction of the rays as determined by Huygens's construction.

Two further verifications must here be mentioned. In the first place, though we shall speak almost exclusively of the propagation of light in transparent dielectrics, a few words may be said about the optical properties of conductors. The simplest assumption concerning the electric current C in a metallic body is expressed by the equation C = QE, where a is the coefficient of conductivity. Combining this with his other formulae. Maxwell found that there must be an absorption of light, a result that can be readily understood since the motion of electricity in a conductor gives rise to a development of heat. But, though Maxwell accounted in this way for the fundamental fact that metals are opaque bodies, there remained a wide divergence between the values of the coefficient of absorption as directly measured and as calculated from the electrical conductivity; but in 1903 it was shown by E. Hagen and H. Rubens that the agreement is very satisfactory in the case of the extreme infra-red rays.

In the second place, the electromagnetic theory requires that a surface struck by a beam of light shall experience a certain pressure. If the beam falls normally on a plane disk, the pressure is normal too; its total amount is given by c-1, if i l, i2 and i 3 are the quantities of energy that are carried forward per unit of time by the incident, the reflected, and the transmitted light.

This result has been quantitatively verified by E. F. Nicholls and G. F. Hull.' Maxwell's predictions have been splendidly confirmed by the experiments of Heinrich Hertz' and others on electromagnetic waves; by diminishing the length of these to the utmost, some physicists have been able to reproduce with them all phenomena of reflection, refraction (single and double), interference, and polarization.

The wave-lengths observed in the aether now has Let the indices *p* and n relate to the two principal cases in which the incident (and, consequently, the reflected) light is polarized in the plane of incidence, or normally to it, and let positive directions *h* and *h be chosen for the disturbance in the incident and for that in the reflected beam, in such a manner that, by a common rotation,* h *and the incident ray prolonged may be made to coincide with* h and the reflected ray.

Mechanical Models of the Electromagnetic Medium

From the results already enumerated, a clear idea can be formed of the difficulties which were encountered in the older form of the wave-theory. Whereas, in Maxwell's theory, longitudinal vibrations are excluded *ab initio* by the solenoidal distribution of the electric current, the elastic-solid theory had to take them into account, unless, as was often done, one made them disappear by supposing them to have a very great velocity of propagation, so that the aether was considered to be practically incompressible. Even on this assumption, however, much in Fresnel's theory remained questionable. Thus George Green,' who was the first to apply the theory of elasticity in an unobjectionable manner, arrived on Fresnel's assumption at a formula for the reflection coefficient *A n* sensibly.

In the theory of double refraction the difficulties are no less serious. As a general rule there are in an anisotropic elastic solid three possible directions of vibration , at right angles to each other, for a given direction of the waves, but none of these lies in the wave-front. In order to make two of them do so and to find Fresnel's form for the wave-surface, new hypotheses are required. On Fresnel's assumption it is even necessary, as was observed by Green, to suppose that in the absence of all vibrations there is already a certain state of pressure in the medium.

If we adhere to Fresnel's assumption, it is indeed scarcely possible to construct an elastic model of the electromagnetic medium. It may be done, however, if the velocities of the particles in the model arc taken to represent the magnetic force H, which, of course, implies that the vibrations of the particles are parallel to the plane of polarization, and that the magnetic energy is represented by the kinetic energy in the model.

Considering further that, in the case of two bodies connected with each other, there is continuity of H in the electromagnetic system, and continuity of the velocity of the particles in the model, it becomes clear that the representation of H by that velocity must be on the same scale in all substances, so that, if E, *n, I* are the displacements of a particle and *g* a universal constant, we may write Hx =gol, Hy - got ' Hz=gat.

By this the magnetic energy per unit of volume becomes zg2 *(OE)* 2 *(mi)* 2 and since this must be the kinetic energy of the elastic medium, the density of the latter must be taken equal to g 2, so that it must be the same in all substances.

THEORIES OF MACCULLAGH

A theory of light in which the elastic aether has a uniform density, and in which the vibrations are supposed to be parallel to the plane of polarization, was developed by Franz Ernst Neumann, 2 who gave the first deduction of the formulas for crystalline reflection.

Like Fresnel, he was, however, obliged to introduce some illegitimate assumptions and simplifications. Here again Green indicated a more rigorous treatment.

James MacCullagh 3 avoided this complication by simply assuming an expression of the form for the potential energy.

He thus established a theory that is perfectly consistent in itself, and may be said to have foreshadowed the electromagnetic theory as regards the form of the equations for transparent bodies.

Lord Kelvin afterwards interpreted MacCullagh's assumption by supposing the only action which is called forth by a displacement to consist in certain couples acting on the elements of volume and proportional to the components If their rotation from the natural position.

He also showed 4 that this "rotational elasticity" can be produced by certain hidden rotations going on in the medium. We cannot dwell here upon other models that have been proposed, and most of which are of rather limited applicability.

A mechanism of a more general kind ought, of course, to be adapted to what is known of the molecular constitution of bodies, and to the highly probable assumption of the perfect permeability for the aether of all ponderable matter, an assumption by which it has been possible to escape from one of the objections raised by Newton.

The possibility of a truly satisfactory model certainly cannot be denied. But it would, in all probability, be extremely complicated.

For this reason many physicists rest content, as regards the free aether, with some such general form of the electromagnetic theory as has been sketched.

OPTICAL PROPERTIES OF PONDERABLE BODIES

Theory of Electrons

If we want to form an: adequate representation of optical phenomena in ponderable bodies, the conceptions of the molecular and atomistic theories naturally suggest themselves. Already, in the elastic theory, it had been imagined that certain material particles are set vibrating by incident waves of light.

These particles had been supposed to be acted on by an elastic force by which they are drawn back towards their positions of equilibrium, so that they can perform free vibrations of their own, and by a resistance that can be represented by terms proportional to the velocity in the equations of motion, and may be physically understood if the vibrations are supposed to be converted in one way or another into a disorderly heat-motion.

In this way it had been found possible to explain the phenomena of dispersion and (selective) absorption, and the connexion between them (anomalous dispersion). 5 These ideas have, been also embodied into the electromagnetic theory.

In its more recent development the extremely small, electrically charged particles, to which the name of " electrons "has been given, and which are supposed to exist in the interior of all bodies, are considered as forming the connecting links between aether and matter, and as determining by their arrangement and their motion all optical phenomena that are not confined to the free aether.6 It has thus become clear why the relations that had been established between optical and electrical properties have been found to hold only in some simple cases.

However, the general boundary conditions seem to require no alteration. For this reason it has been possible, for example, to establish a satisfactory theory of metallic reflection, though the propagation of light in the interior of a metal is only imperfectly understood.

One of the fundamental propositions of the theory of electrons is that an electron becomes a centre of radiation whenever its velocity changes either in direction or in magnitude.

Thus the production of Röntgen rays, regarded as consisting of very short and irregular electromagnetic impulses, is traced to the impacts of the electrons of the cathode-rays against the anticathode, and the lines of an emission spectrum indicate the existence in the radiating body of as many kinds of regular vibrations, the knowledge of which is the ultimate object of our investigations about the structure of the spectra.

The shifting of the lines caused, according to Doppler's law, by a motion of the source of light, may easily be accounted for, as only general principles are involved in the explanation. To a certain extent we can also elucidate the changes in the emission that are observed when the radiating source is exposed to external magnetic foooptics) .

Various Kinds of Light-motion

- If the disturbance is represented by P z = o, P = a cos *(nt - kx+f), 'P Z = a'cos (nt - + f)*, so that the end of the vector P describes an ellipse in a plane perpendicular to the direction of propagation, the light is said to be elliptically, or in special cases circularly, polarized. Light of this kind can be dissolved in many different ways into plane polarized components.
  There are cases in which plane waves must be elliptically or circularly polarized in order to show the simple propagation of phase that is expressed by formulae. Instances of this kind occur in bodies having the property of rotating the plane of polarization, either on account of their constitution, or under the influence of a magnetic field. For a given direction of the wave-front there are in general two kinds of elliptic vibrations, each having a definite form, orientation, and direction of motion, and a determinate velocity of propagation. All that has been said about Huygens's construction applies to these cases.

- In a perfect spectroscope a sharp line would only be observed if an endless regular succession of simple harmonic vibrations were admitted into the instrument. In any other case the light will occupy a certain extent in the spectrum, and in order to determine its distribution we have to decompose into simple harmonic functions of the time the components of the disturbance, at a point of the slit for instance. This may be done by means of Fourier's theorem.

  An extreme case is that of the unpolarized light emitted by incandescent solid bodies, consisting of disturbances whose variations are highly irregular, and giving a continuous spectrum. But even with what is commonly called homogeneous light, no perfectly sharp line will be seen. There is no source of light in which the vibrations of the particles remain for ever undisturbed, and a particle will never emit an endless succession of uninterrupted vibrations, but at best a series of vibrations whose form, phase and intensity are changed at irregular intervals. The result must be a broadening of the spectral line.

  In cases of this kind one must distinguish between the velocity of propagation of the phase of regular vibrations and the velocity with which the said changes travel onward.
- In a train of plane waves of definite frequency the disturbance is represented by means of goniometric functions of the time and the coordinates. Since the fundamental equations are linear, there are also solutions in which one or more of the coordinates occur in an exponential function. These solutions are of interest because the motions corresponding to them are widely different from those of which we have thus far spoken. If, for example, the formulae contain the factor e - *zcos* (nt - sy +1) *with the positive constant r, the disturbance is no longer periodic with respect to x, but steadily diminishes as x increases. A state*

of things of this kind, in which the vibrations rapidly die away as we leave the surface, exists in the air adjacent to the face of a glass prism by which a beam of light is totally reflected. It furnishes us an explanation of Newton's experiment mentioned.

Velocity of Light

The fact that light is propagated with a definite speed was first brought out by Ole Roemer at Paris, in 1676, through observations of the eclipses of Jupiter's satellites, made in different relative positions of the Earth and Jupiter in their respective orbits. It is possible in this way to determine the time required for light to pass across the orbit of the earth. The dimensions of this orbit, or the distance of the sun, being taken as known, the actual speed of light could be computed. Since this computation requires a knowledge of the sun's distance, which has not yet been acquired with certainty, the actual speed is now determined by experiments made on the earth's surface.

Were it possible by any system of signals to compare with absolute precision the times at two different stations, the speed could be determined by finding how long was required for light to pass from one station to another at the greatest visible distance. But this is impracticable, because no natural agent is under our control by which a signal could be communicated with a greater velocity than that of light. It is therefore necessary to reflect a ray back to the point of observation and to determine the time which the light requires to go and come. Two systems have been devised for this purpose. One is that of Fizeau, in which the vital appliance is a rapidly revolving toothed wheel; the other is that of Foucault, in which the corresponding appliance is a mirror revolving on an axis in, or parallel to, its own plane.

The principle underlying Fizeau's method is shown in the accompanying the course of a ray of light which, emanating from a luminous point L, strikes the plane surface of a plate of glass M at an angle of about 45° A fraction of the light is reflected from the two surfaces of the glass to a distant reflector

R, the plane of which is at right angles to the course of the ray. The latter is thus reflected back on its own course and, passing through the glass M on its return, reaches a point E behind the glass. An observer with his eye at E iooking through the glass sees the return ray as a distant luminous point in the reflector R, after the Nam light has passed over the course in both directions trance from M nearly equal to its focal length. The function of this appliance is to render the diverging rays, shown by the dotted lines, nearly parallel, in order that more light may reach R and be thrown back again. But the principle may be conceived without respect to the telescope, all the rays being ignored except the central one, which passes over the course we have described.

The luminous point between two of the teeth at K. Now, conceive that the wheel is set in revolution. The ray is then interrupted as every tooth passes, so that what is sent out is a succession of flashes. Conceive that the speed of the mirror is such that while the flash is going to the distant mirror and returning again, each tooth of the wheel takes the place of an opening between the teeth.

Then each flash sent out will, on its return, be intercepted by the adjacent tooth, and will therefore become invisible. If the speed be now doubled, so that the teeth pass at intervals equal to the time required for the light to go and come, each flash sent through an opening will return through the adjacent opening, and will therefore be seen with full brightness. If the speed be continuously increased the result will be successive disappearances and reappearances of the light, according as a tooth is or is not interposed when the ray reaches the apparatus on its return. The computation of the time of passage and return is then very simple. The speed of the wheel being known, the number of teeth passing in one second can be computed. The order of the disappearance, or the number of teeth which have passed while the light is going and coming, being also determined in each case, the interval of time is computed by a simple formula.

The most elaborate determination yet 'made by Fizeau's method was that of Cornu. The station of observation was at the Paris Observatory. The distant reflector, a telescope with a reflector at its focus, was at Montlhery, distant 22,910 metres from the toothed wheel. Of the wheels most used one had 150 teeth, and was 35 millimetres in diameter; the other had 200 teeth, with a diameter of 45 mm.

The highest speed attained was about 900 revolutions per second. At this speed, 135,000 (or 180,000) teeth would pass per second, and about 20 (or 28) would pass while the light was going and coming. But the actual speed attained was generally less than this.

The definitive result derived by Cornu from the entire series of experiments was 300,400 kilometres per second. Further details of this work need not be set forth because the method is in several ways deficient in precision. The eclipses and subsequent reappearances of the light taking place gradually, it is impossible to fix with entire precision upon the moment of complete eclipse. The speed of the wheel is continually varying, and it is impossible to determine with precision what it was at the instant of an eclipse.

The defect would be lessened were the speed of the toothed wheel placed under control of the observer who, by action in one direction or the other, could continually check or accelerate it, so as to keep the return point of light at the required phase of brightness.

If the phase of complete extinction is chosen for this purpose a definite result cannot be reached; but by choosing the moment when the light is of a certain definite brightness, before or after an eclipse, the observer will know at each instant whether the speed should be accelerated or retarded, and can act accordingly. The nearly constant speed through as long a period as is deemed necessary would then be found by dividing the entire number of revolutions of the wheel by the time through which the light was kept constant. But even with these improvements, which were not actually tried by Cornu, the estimate of the brightness on which the whole result depends would necessarily be uncertain.

The outcome is that, although Cornu's discussion of his experiments is a model in the care taken to determine so far as practicable every source of error, his definitive result is shown by other determinations to have been too great by about a t°u part of its whole amount.

In Foucault's determination the measures were not made upon a luminous point, but upon a reticule, the image of which could not be seen unless the reflector was quite near the revolving mirror. Indeed the whole apparatus was contained in his laboratory. The effective distance was increased by using several reflectors; but the entire course of the ray measured only 20 metres. The result reached by Foucault for the velocity of light was 298,000 kilometres per second. The first marked advance on Foucault's determination was made by Albert A. Michelson, then a young officer on duty at the U.S. Naval Academy, Annapolis.

The improvement *Michelson* consisted in using the image of a slit through which the rays of the sun passed after reflection from a heliostat. In this way it was found possible to see the image of the slit reflected from the distant mirror when the latter was nearly 600 metres from the station of observation.

It will be seen that the revolving mirror is here interposed between the lens and its focus. It was driven by an air turbine, the blast of which was under the control of the observer, so that it could be kept at any required speed. The speed was determined by the vibrations of two tuning forks. One of these was an electric fork, making about 120 vibrations per second, with which the mirror was kept in unison by a system of rays reflected from it and the fork. The speed of this fork was determined by comparison with a freely vibrating fork from time to time. The speed of the revolving mirror was generally about 275 turns per second, and the deflection of the image of the slit about 112.5 mm.

In order that the disturbances of the return image due to the passage of the ray through more than 7 km. of air might be reduced to a minimum, an ordinary telescope of the " broken back " form was used to send the ray to the revolving mirror. The speed of the mirror was, as in Michelson's experiments, completely under control of the observer, so that by drawing one or the other of two cords held in the hand the return image could be kept in any required position.

In making each measure the receiving telescope hereafter described was placed in a fixed position and during the " run " the image was kept as nearly as practicable upon a vertical thread passing through its focus. A " run " generally lasted about two minutes, during which time the mirror commonly made between 25,000 and 30,000 revolutions. The speed per second was found by dividing the entire number of revolutions by the number of seconds in the "run"

The extreme deviations between the times of transmission of the light, as derived from any two runs, never approached to the thousandth part of its entire amount. The average deviation from the mean was indeed less than Pa 0 part of the whole. To avoid the injurious effect of the directly reflected flash, as well as to render unnecessary a comparison between the directions of the outgoing and the return ray. a second telescope, turning horizontally on an axis coincident with that of the revolving mirror, was used to receive the return ray after reflection.

This required the use of an elongated mirror of which the upper half of the surface reflected the outgoing ray, and the lower other half received and reflected the ray on its return. On this system it was not necessary to incline the mirror in order to avoid the direct reflection of the return ray. The greatest advantage of this system was that the revolving mirror could be turned in either direction without break of continuity, so that the angular measures were made between the directions of the return ray after reflection when the mirror moved in opposite directions. In this way the speed of the mirror was as good as doubled, and the possible constant errors inherent in the reference to a fixed direction for the sending telescope were eliminated.

The revolving mirror was a rectangular prism M of steel, 3 in. high and 12 in. on a side in cross section, which was driven A by a blast of air acting on two fanwheels, one at the top, the other at the bottom of the mirror. NPO is the object-end of the fixed sending telescope the rays passing through it' being reflected to the mirror by a prism P. The receiving telescope ABO is straight, and has its objective under O. It was attached to a frame which could turn around the same axis as the mirror. The angle through which it moved was measured by a divided arc immediately below its eye-piece. The position AB is that for receiving the ray during a rotation of the mirror in the anti-clockwise direction; the position A'B' that for a clockwise rotation.

In these measures the observing station was at Fort Myer, on a hill above the west bank of the Potomac river. The distant reflector was first placed in the grounds of the Naval Observatory, at a distance of 2551 metres. But the definitive measures were made with the reflector at the base of the Washington monument, 3721 metres distant. The revolving mirror was of nickel-plated steel, polished on all four vertical sides.

Thus four reflections of the ray were received during each turn of the mirror, which would be coincident were the form of the mirror invariable. During the preliminary series of measures it was found that two images of the return ray were sometimes formed, which would result in two different conclusions as to the velocity of light, according as one or the other was observed. The only explanation of this defect which presented itself was a tortional vibration of the revolving mirror, coinciding in period with that of revolution, but it was first thought that the effect was only occasional.

In the summer of 1881 the distant reflector was removed from the Observatory to the Monument station. Six measures made in August and September showed a systematic deviation of +67 km. per second from the result of the Observatory series. This difference led to measures for eliminating the defect from which it was supposed to arise. The pivots of the mirror were reground, and a change made in the arrangement, which would permit of the effect of the vibration being determined and eliminated.

This consisted in making the relative position of the sending and receiving telescopes interchangeable. In this way, if the measured deflection was too great in one position of the telescopes, it would be too small by an equal amount in the reverse position.

As a matter of fact, when the definitive measures were made, it was found that with the improved pivots the mean result was the same in the two positions. But the new result differed systematically from both the former ones. Thirteen measures were made from the Monument in the summer of 1882, the results of which will first be stated in the form of the time required by the ray to go and come.

So far as could be determined from the discordance of the separate measures, the mean error of Newcomb's result would be less than Da km. But making allowance for the various sources of systematic error the actual probable error was estimated at =30 km. It seems remarkable that since these determinations were made, a period during which great improvements have become possible in every part of the apparatus, no complete redetermination of this fundamental physical constant has been carried out.

The experimental measures thus far cited have been primarily those of the velocity of light in air, the reduction to a vacuum being derived from theory alone. The fundamental constant at the basis of the whole theory is the speed of light in a vacuum, such as the celestial spaces. The question of the relation between the velocity in vacuo, and in a transparent medium of any sort, belongs to the domain of physical optics. Referring to the preceding section for the principles at play we shall in the present part of the article confine ourselves to the experimental results. With the theory of the effect of a transparent medium is associated that of the possible differences in the speed of light of different colours.

The question whether the speed of light in vacuo varies with its wave-length seems to be settled with entire certainty by observations of variable stars. These are situated at different distances, some being so far that light must be several centuries in reaching us from them.

Were there any difference in the speed of light of various colours it would be shown by a change in the colour of the star as its light waxed and waned. The light of greatest speed preceding that of lesser speed would, when emanated during the rising phase, impress its own colour on that which it overtook. The slower light would predominate during the falling phase. If there were a difference of Io minutes in the time at which light from the two ends of the visible spectrum arrived, it would be shown by this test. As not the slightest effect of the kind has ever been seen, it seems certain that the difference, if any, cannot imate to T. 0 0.0 0 0 part rt of the entire speed.

The approx case is different when light passes through a refracting medium. It is a theoretical result of the undulatory theory of light that its velocity in such a medium is inversely proportional to the refractive index of the medium. This being different for different colours, we must expects a corresponding difference in the velocity.

Foucault and Michelson have tested these results of the undulatory theory by comparing the time required for a ray of light to pass through a tube filled with a refracting medium, and through air. Foucault thus found, in a general way, that there actually was a retardation; but his observations took account only of the mean retardation of light of all the wavelengths, which he found to correspond with the undulatory theory. Michelson went further by determining the retardation of light of various wave-lengths in carbon bisulphide. The comparison of red and blue light was made differentially. The colours selected were of wave-length about o 62 for red and 0.49 for blue. Putting Vr and Vb for the speeds of red and blue light respectively in bisulphide of carbon, the mean result compares with theory as follows: Observed value of the ratio V r, Theoretical value (Verdet).. 1.025 This agreement may be regarded as perfect. It shows that the divergence of the speed of yellow light in the medium from theory, as found above, holds through the entirespectrum.

The excess of the retardation above that resulting from theory is probably due to a difference between " wave-speed" and "group-speed" pointed out by Rayleigh.

1.0245 But when a flash of light like that measured passes through a refracting medium, the front waves of the flash are continually dying away, and the place of each is taken by the wave following. A familiar case of this sort is seen when a stone is thrown into a pond. The front waves die out one at a time, to be followed by others, each of which goes further than its predecessor, while new waves are formed in the rear.

Larger waves in the middle, moves as a whole more slowly than do the individual waves. When the speed of light is measured the result is not the wave-speed as above defined, but something less, because the result depends on the time of the group passing through the medium. This lower speed is called the groupvelocity of light. In a vacuum there is no dying out of the waves, so that the group-speed and the wave-speed are identical. From Michelson's experiments it would follow that the retardation was about 1/14 of the whole speed. This would indicate that in carbon bisulphide each individual light wave forming the front of a moving ray dies out in a space of about 15 wavelengths.

WAVE-SURFACE

After having found the values of v for a particular frequency and different directions of the wavenormal, a very instructive graphical representation can be employed. Let ON be a line in any direction, drawn from a fixed point 0, OA a length along this line equal to the velocity v of waves having ON for their normal, or, more generally, OA, OA′, &e., lengths equal to the velocities v, v', &c., which such waves have according to their direction of vibration, Q, Q′, &c., planes perpendicular to ON through A, A, &c. Let this construction be repeated for all directions of ON, and let W be the surface that is touched by all the planes Q, Q′, &c.

It is clear that if this surface, which is called the " wave-surface," is known, the velocity of propagation of plane waves of any chosen direction is given by the length of the perpendicular from the centre 0 on a tangent plane in the given direction.

It must be kept in mind that, in general, each tangent plane corresponds to one definite direction of vibration. If this direction is assigned in each point of the wavesurface, the diagram contains all the information which we can desire concerning the propagation of plane waves of the frequency that has been chosen.

The plane Q employed in the above construction is the position after unit of time of a wave-front perpendicular to ON and originally passing through the point O. The surface W itself is often considered as the locus of all points that are reached in unit of time by a disturbance starting from 0 and spreading towards all sides.

Admitting the validity of this view, we can determine in a similar way the locus of the points reached in some infinitely short time *dt*, the wavesurface, as we may say, or the " elementary wave," corresponding to this time. It is similar to W, all dimensions of the latter surface being multiplied by *dt*. It may be noticed that in a heterogeneous medium a wave of this kind has the same form as if the properties of matter existing at its centre extended over a finite space.

Theory of Huygens

Huygens was the first to show that the explanation of optical phenomena may be made to depend on the wave-surface, not only in isotropic bodies, in which it has a spherical form, but also in crystals, for one of which deduced the form of the surface from the observed double refraction. In his argument Huygens availed himself of the following principle that is justly named after him:

Any point that is reached by a wave of light becomes a new centre of radiation from which the disturbance is propagated towards all sides. On this basis he determined the progress of light-waves by a construction which, under a restriction to be applied to waves of any form and to all kinds of transparent media.

Let obe the surface (wave-front) to which a definite phase of vibration has advanced at a certain time *t, dt* an infinitely small increment of time, and let an elementary wave corresponding to this interval be described around each point P of a. Then the envelope v′ of all these elementary waves is the surface reached by the phase in question at the time *t+dt,* and by repeating the construction all successive positions of the wave-front can be found.

Huygens also considered the propagation of waves that are laterally limited, by having passed, for example, through an opening in an opaque screen. If, in the first wave-front *a*, the disturbance exists only in a certain part bounded by the contour *s*, we can confine ourselves to the elementary waves around the points of that part, and to a portion of the new wave-front a′ whose boundary passes through the points where a′ touches the elementary waves having their centres on *s*.

Taking for granted Huygens's assumption that a sensible disturbance is only found in those places where the elementary waves are touched by the new wave-front, it may be inferred that the lateral limits of the beam of light are determined by lines, each element of which joins the centre P of an elementary wave with its point of contact P′ with the next wave-front. To lines of this kind, whose course can be made visible by using narrow pencils of light, the name of " rays " is to be given in the wave-theory.

The disturbance may be conceived to travel along them with a velocity $u = PP'/dt$, which is therefore called the " ray-velocity." The construction shows that, corresponding to each direction of the wave-front (with a determinate direction of vibration), there is a definite direction and a definite velocity of the ray.

Chapter 2

Geometric Optics

Optics deals with the propagation of light through transparent media, and its interaction with mirrors, lenses, slits, *etc*. Optical effects can be divided into two broad classes. Firstly, those which can be explained without reference to the fact that light is fundamentally a wave phenomenon, and, secondly, those which can only be explained on the basis that light is a wave phenomenon. Let us, for the moment, consider the former class of effects. It might seem somewhat surprising that any optical effects at all can be accounted for without reference to waves.

Light really is a wave phenomenon. It turns out, however, that wave effects are only crucially important when the wavelength of the wave is either comparable to, or much larger than, the size of the objects with which it interacts. When the wavelength of the wave becomes much smaller than the size of the objects with which it interacts then the interactions can be accounted for in a very simple geometric manner. Since the wavelength of visible light is only of order a micron, it is very easy to find situations in which its wavelength is very much smaller than the size of the objects with which it interacts. Thus, "wave-less" optics, which is usually called *geometric optics*, has a very wide range of applications.

In geometric optics, light is treated as a set of *rays*, emanating from a source, which propagate through transparent media according to a set of *three* simple laws. The first law is the *law of rectilinear propagation*, which states that light rays propagating through a homogeneous transparent medium do so in straight-lines. The second law is the *law of reflection,* which governs the interaction of light rays with conducting surfaces (*e.g.,* metallic mirrors). The third law is the *law of refraction,* which governs the behaviour of light rays as they traverse a sharp boundary between two different transparent media (*e.g.,* air and glass).

HISTORY OF GEOMETRIC OPTICS

Let us first consider the law of *rectilinear propagation.* The earliest surviving optical treatise, Euclid's *Catoptrics,* recognized that light travels in straight-lines in homogeneous media. However, following the teachings of Plato, Euclid thought that light rays emanate from the eye, and intercept external objects, which are thereby "seen" by the observer. The ancient Greeks also thought that the speed with which light rays emerge from the eye is very high, if not infinite. After all, they argued, an observer with his eyes closed can open them and immediately see the distant stars.

Hero of Alexandria, in his *Catoptrics* (first century BC), also maintained that light travels with infinite speed. His argument was by analogy with the free fall of objects. If we throw an object horizontally with a relatively small velocity then it manifestly does not move in a straight-line. However, if we throw an object horizontally with a relatively large velocity then it appears to move in a straight-line to begin with, but eventually deviates from this path. The larger the velocity with which the object is thrown, the longer the initial period of apparent rectilinear motion.

Hero reasoned that if an object were thrown with an infinite velocity then it would move in a straight-line forever. Thus, light, which travels in a straight-line, must move with an infinite velocity. The erroneous idea that light travels with an *infinite* velocity persisted until 1676, when the Danish astronomer Olaf Römer demonstrated that light must have a *finite* velocity, using his timings of the successive eclipses of the satellites of Jupiter, as they passed into the shadow of the planet.

Ptolemy formulated a very inaccurate version of the law of refraction, which only works when the light rays are almost normally incident on the interface in question. Despite its obvious inaccuracy, Ptolemy's theory of refraction persisted for nearly 1500 years.

The true law of refraction was discovered empirically by the Dutch mathematician Willebrord Snell in 1621. However, the French philosopher René Descartes was the first to publish, in his *La Dioptrique* the now familiar formulation of the law of refraction in terms of sines. Although there was much controversy at the time regarding plagiarism, Descartes was apparently unaware of Snell's work. Thus, in English speaking countries the law of refraction is called "Snell's law", but in French speaking countries it is called "Descartes' law".

In 1658, the French mathematician Pierre de Fermat demonstrated that all three of the laws of geometric optics can be accounted for on the assumption that light always travels between two points on the path which takes the *least time* (or, more rigorously, the extremal time). Fermat's ideas were an extension of those of Hero of Alexandria. Fermat's (correct) derivation of the law of refraction depended crucially on his (correct) assumption that light travels *more slowly* in dense media than it does in air. Unfortunately, many famous scientists, including Newton, maintained that light travels *faster* in dense media than it does in air.

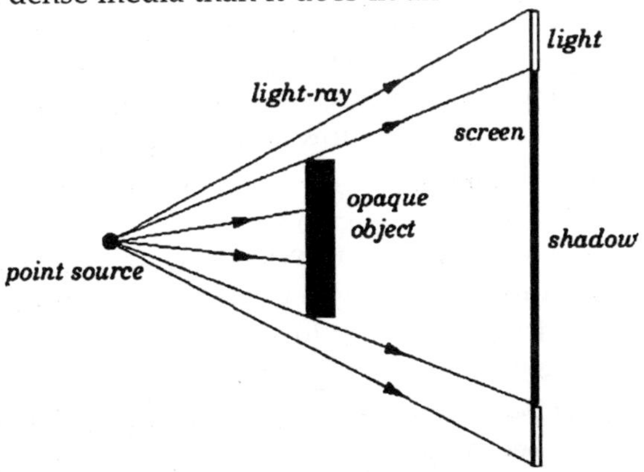

Fig. An Opaque Object Illuminated by a Point Light Source.

This erroneous idea held up progress in optics for over one hundred years, and was not conclusively disproved until the mid-nineteenth century. Incidentally, Fermat's principle of least time can only be justified using wave theory.

LAW OF GEOMETRIC PROPAGATION

According to geometric optics, an opaque object illuminated by a point source of light casts a sharp shadow whose dimensions can be calculated using geometry. The method of calculation is very straightforward. The source emits *light-rays* uniformly in all directions. These rays can be represented as straight lines radiating from the source. The light-rays propagate away from the source until they encounter an opaque object, at which point they stop.

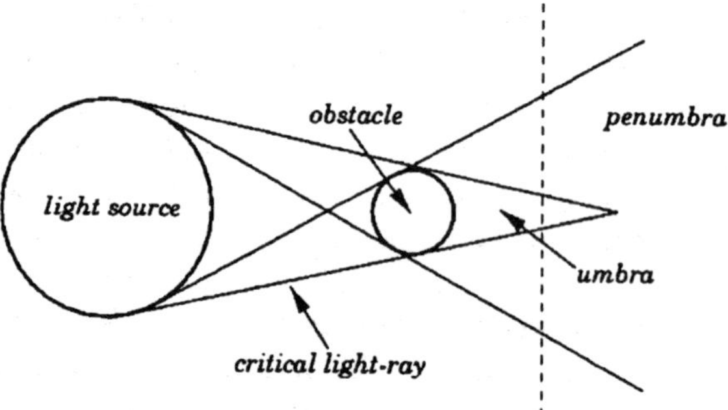

Fig. An opaque Object Illuminated by an Extended Light Source.

For an extended light source, each element of the source emits light-rays, just like a point source. Rays emanating from different elements of the source are assumed not to interfere with one another. The shadow cast by an opaque sphere illuminated by a spherical light source is calculated using a small number of critical light-rays.

The shadow consists of a perfectly black disk called the *umbra,* surrounded by a ring of gradually diminishing darkness called the *penumbra.* In the umbra, *all* of the light-rays emitted by the source are blocked by the opaque sphere, whereas in the penumbra only *some* of the rays emitted by the source are blocked by the sphere. As was well-known to the ancient Greeks, if the light-source represents the Sun, and the opaque sphere the Moon, then at a point on the Earth's surface which is situated inside the umbra the Sun is totally eclipsed, whereas at a point on the Earth's surface which is situated in the penumbra the Sun is only partially eclipsed.

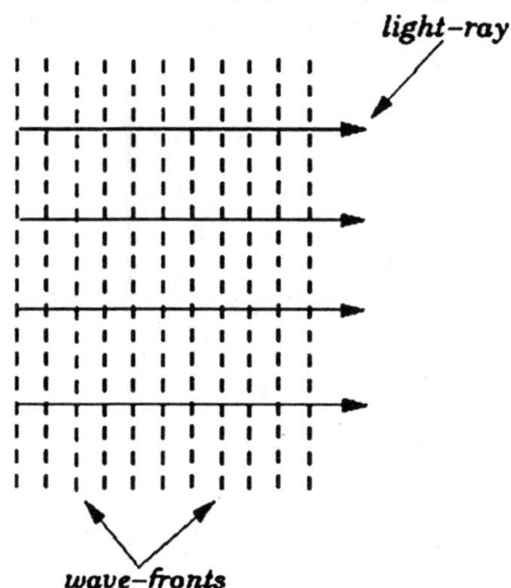

Fig. Relationship between Wave-Fronts and Light-Rays.

In the wave picture of light, a *wave-front* is defined as a surface joining all adjacent points on a wave that have the same phase (*e.g.*, all maxima, or minima, of the electric field). A light-ray is simply a line which runs perpendicular to the wave-fronts at all points along the path of the wave.

Thus, the law of rectilinear propagation of light-rays also specifies how wave-fronts propagate through homogeneous media. Of course, this law is only valid in the limit where the wavelength of the wave is much smaller than the dimensions of any obstacles which it encounters.

REFLECTION AND REFRACTION

Most of what we need to know about geometrical optics can be summarized in two rules, the laws of reflection and refraction. These rules may both be inferred by considering what happens when a plane wave segment impinges on a flat surface. If the surface is polished metal, the wave is *reflected,* whereas if the surface is an interface between two transparent media with differing indices of refraction, the wave is partially reflected and partially *refracted*. Reflection means that the wave is turned back into the half-space from which it came, while refraction means that it passes through the interface, acquiring a different direction of motion from that which it had before reaching the interface.

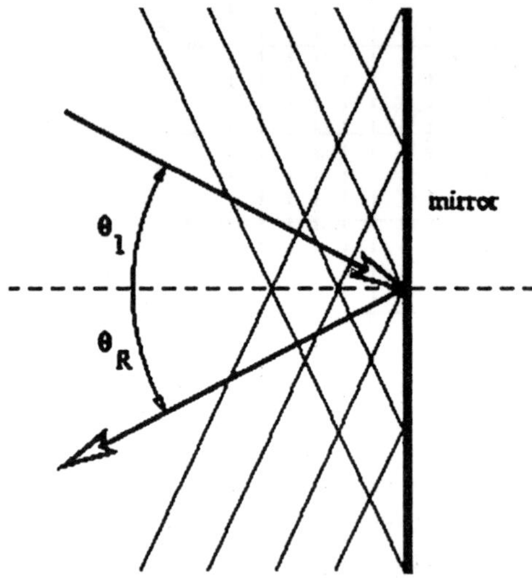

Fig. Sketch Showing the Reflection of a Wave from A Plane Mirror. The Law of Reflection States that $\theta_I = \theta_R$.

The wave vector and wave front of a wave being reflected from a plane mirror. The angles of incidence, θ_I, and reflection, θ_R, are defined to be the angles between the incoming and outgoing wave vectors respectively and the line normal to the mirror. The law of reflection states that $\theta_R = \theta_I$. This is a consequence of the need for the incoming and outgoing wave fronts to be in phase with each other all along the mirror surface.

This plus the equality of the incoming and outgoing wavelengths is sufficient to insure the above result. Since $n_R > n_I$, the speed of light in the right-hand medium is less than in the left-hand medium. (Recall that the speed of light in a medium with refractive index n is $c_{\text{medium}} = c_{vac}/n$.) The frequency of the wave packet doesn't change as it passes through the interface, so the wavelength of the light on the right side is less than the wavelength on the left side. The side AC is equal to the side BC times $\sin(\theta_I)$.

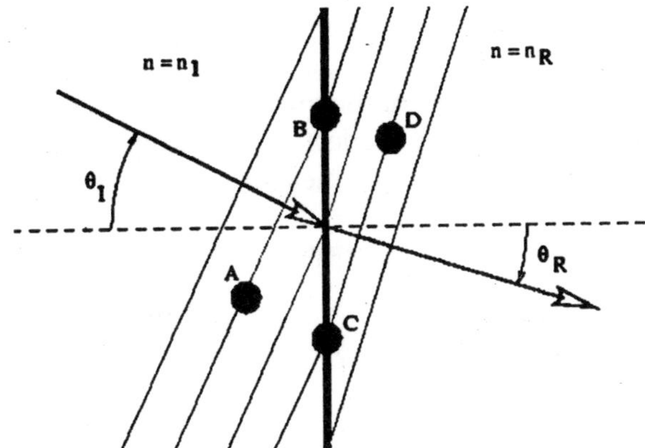

Fig. Sketch Showing the Refraction of A Wave from an Interface between two Dielectric Media with $n_2 > n_1$.

However, AC is also equal to $2\lambda_I$, or twice the wavelength of the wave to the left of the interface. Similar reasoning shows that $2l_R$, twice the wavelength to the right of the interface, equals BC times sin (θ_R). Since the interval BC is common to both triangles

$$\frac{\lambda_I}{\lambda_R} = \frac{\sin(\theta_I)}{\sin(\theta_R)}.$$

Since $\lambda_I = c_I T = c_{vac} T/n_I$ and $\lambda_R = C_R T = c_{vac} \mathrm{T}/n_R$ where C_I and C_R are the wave speeds to the left and right of the interface, C_{vac} is the speed of light in a vacuum, and T is the (common) period, we can easily recast the above equation in the form

$$n_I \sin(\theta_I) = n_R \sin(\theta_R).$$

This is called *Snell's law,* and it governs how a ray of light bends as it passes through a discontinuity in the index of refraction. The angle θ_I is called the incident angle and θ_R is called the refracted angle. Notice that these angles are measured from the normal to the surface, not the tangent.

Law of Reflection

The law of reflection governs the reflection of light-rays off smooth conducting surfaces, such as polished metal or metal-coated glass mirrors.

Consider a light-ray incident on a plane mirror. The law of reflection states that the incident ray, the reflected ray, and the normal to the surface of the mirror all lie in the *same plane.* Furthermore, the angle of reflection r is *equal* to the angle of incidence i. Both angles are measured with respect to the normal to the mirror.

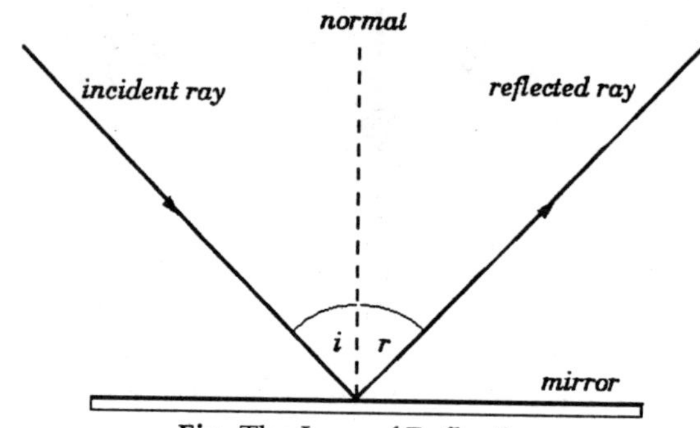

Fig. The Law of Reflection

The law of reflection also holds for non-plane mirrors, provided that the normal at any point on the mirror is understood to be the outward pointing normal to the local tangent plane of the mirror at that point. For rough surfaces, the law of reflection remains valid.

It predicts that rays incident at slightly different points on the surface are reflected in completely different directions, because the normal to a rough surface varies in direction very strongly from point to point on the surface. This type of reflection is called *diffuse reflection,* and is what enables us to see non-shiny objects.

Law of Refraction

The law of refraction, which is generally known as *Snell's law,* governs the behaviour of light-rays as they propagate across a sharp interface between two transparent dielectric media. Consider a light-ray incident on a plane interface between two transparent dielectric media. The law of refraction states that the incident ray, the refracted ray, and the normal to the interface, all lie in the *same plane.*

$$n_1 \sin\theta_1 = n_2 \sin\theta_2,$$

where θ_1 is the angle subtended between the incident ray and the normal to the interface, and θ_2 is the angle subtended between the refracted ray and the normal to the interface. The quantities n_1 and n_2 are termed the *refractive indices* of media 1 and 2, respectively.

Thus, the law of refraction predicts that a light-ray always deviates more towards the normal in the optically denser medium: *i.e.,* the medium with the higher refractive index. Note that $n_2 > n_1$ in the figure. The law of refraction also holds for non-planar interfaces, provided that the normal to the interface at any given point is understood to be the normal to the local tangent plane of the interface at that point.

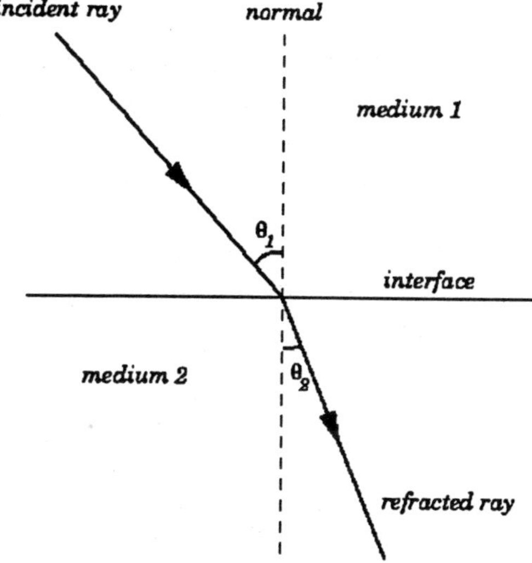

Fig. The Law of Refraction.

By definition, the refractive index n of a dielectric medium of dielectric constant K is given by

$$n = \sqrt{K}.$$

The refractive indices of some common materials (for yellow light of wavelength l = 589 nm).

The law of refraction follows directly from the fact that the speed v with which light propagates through a dielectric medium is *inversely proportional* to the refractive index of the medium.

Table. Refractive Indices of Some Common Materials at l = 589 nm.

Material	n
Air (STP)	1.00029
Water	1.33
Ice	1.31
Glass:	
Light flint	1.58
Heavy flint	1.65
Heaviest flint	1.89
Diamond	2.42

In fact,

$$v = \frac{c}{n},$$

where c is the speed of light in a vacuum. Consider two parallel light-rays, a and b, incident at an angle θ_1 with respect to the normal to the interface between two dielectric media, 1 and 2. Let the refractive indices of the two media be n_1 and n_2 respectively, with $n_2 > n_1$.

It is clear from ray b must move from point B to point Q, in medium 1, in the same time interval, Δt, in which ray a moves between points A and P, in medium 2. Now, the speed of light in medium 1 is $v_1 = c/n_1$, whereas the speed of light in medium 2 is $v_2 = c/n_2$. It follows that the length BQ is given by $v_1\,\Delta t$, whereas the length AP is given by $v_2\,\Delta t$. By trigonometry,

$$\sin\theta_1 = \frac{AP}{AQ} = \frac{v_1\Delta t}{AQ},$$

and

$$\sin\theta_2 = \frac{AP}{AQ} = \frac{v_2\Delta t}{AQ}.$$

Hence,

$$\frac{\sin\theta_1}{\sin\theta_2} = \frac{v_1}{v_2} = \frac{n_2}{n_1},$$

which can be rearranged to give Snell's law. Note that the lines AB and PQ represent wave-fronts in media 1 and 2, respectively, and, therefore, cross rays a and b at right-angles.

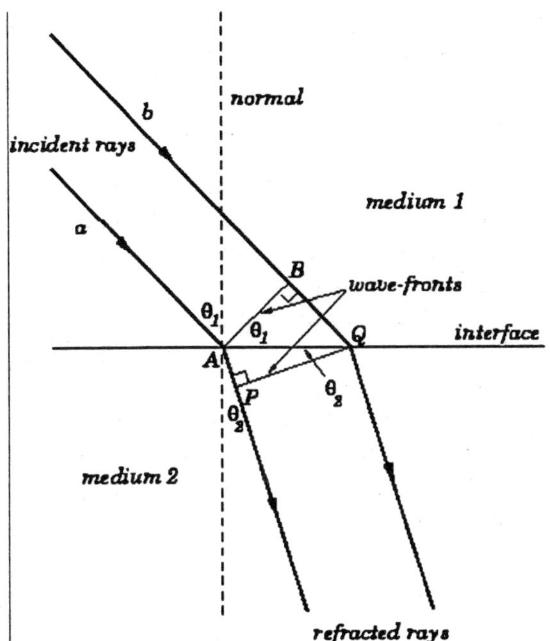

Fig. Derivation of Snell's Law.

When light passes from one dielectric medium to another its velocity v changes, but its frequency f remains *unchanged*. Since, $v = f\lambda$ for all waves, where λ is the wavelength, it follows that the wavelength of light must also change as it crosses an interface between two different media. Suppose that light propagates from medium 1 to medium 2. Let n_1 and n_2 be the refractive indices of the two media, respectively. The ratio of the wave-lengths in the two media is given by

$$\frac{\lambda_2}{\lambda_1} = \frac{v_2 / f}{v_1 / f} = \frac{v_2}{v_1} = \frac{n_1}{n_2}.$$

Thus, as light moves from air to glass its wavelength *decreases*.

TOTAL INTERNAL REFLECTION

An interesting effect known as *total internal reflection* can occur when light attempts to move from a medium having a given refractive index to a medium having a *lower* refractive index. Suppose that light crosses an interface from medium 1 to medium 2, where $n_2 < n_1$. According to Snell's law,

$$\sin\theta_2 = \frac{n_1}{n_2}\sin\theta_1.$$

Since $n_1/n_2 > 1$, it follows that $\theta_2 > \theta_1$. For relatively small angles of incidence, part of the light is refracted into the less optically dense medium, and part is reflected (there is always some reflection at an interface). When the angle of incidence θ_1 is such that the angle of refraction $\theta_2 = 90°$, the refracted ray runs along the interface between the two media. This particular angle of incidence is called the *critical angle*, θ_c. For $\theta_1 > \theta_c$, there is *no* refracted ray. This effect is called *total internal reflection*, and occurs whenever the angle of incidence exceeds the critical angle. Now when $\theta_1 = \theta_c$, we have $\theta_2 = 90°$, and so $\sin\theta_2 = 1$. It follows from Eq. that

$$\sin\theta_c = \frac{n_2}{n_1}.$$

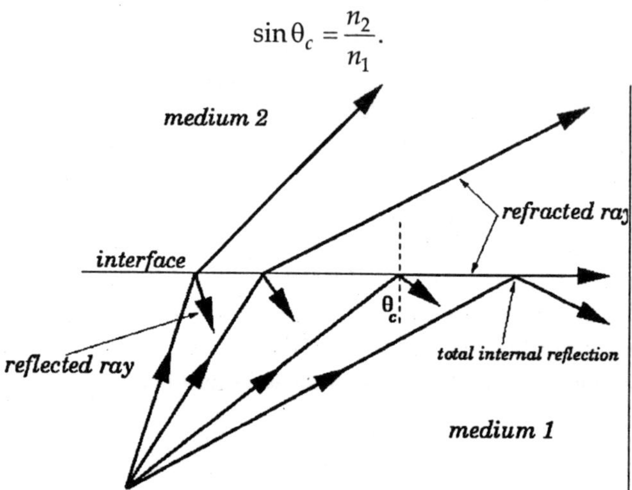

Fig. Total Internal Reflection.

Consider a fish swimming in a clear pond. If the fish looks upwards it sees the sky, but if it looks at too large an angle to the vertical it sees the bottom of the pond reflected on the surface of the water. The critical angle to the vertical at which the fish first sees the reflection of the bottom of the pond is, of course, equal to the critical angle θ_c for total internal reflection at an air-water interface. From Eq. this critical angle is given by

$$\theta_c = \sin^{-1}(1.00/1.33) = 48.8°,$$

since the refractive index of air is approximately unity, and the refractive index of water is 1.33.

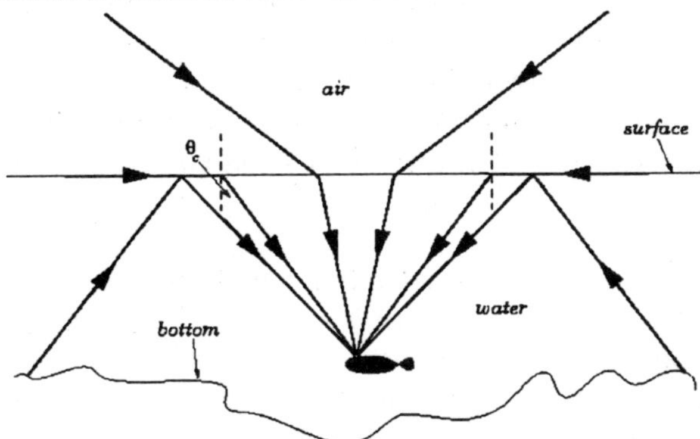

Fig. A fish's Eye View.

When total internal reflection occurs at an interface the interface in question acts as a *perfect reflector*. This allows 45° crown glass prisms to be used, in place of mirrors, to reflect light in binoculars. The angles of incidence on the sides of the prism are all 45°, which is greater than the critical angle 41° for crown glass (at an air-glass interface).

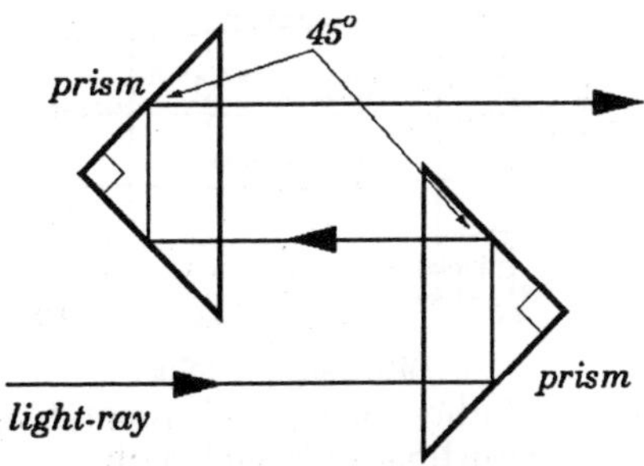

Fig. Arrangement of Prisms Used in Binoculars.

Diamonds, for which $n = 2.42$, have a critical angle θ_c which is only 24°. The facets on a diamond are cut in such a manner that much of the incident light on the diamond is reflected many times by successive total internal reflections before it escapes. This effect gives rise to the characteristic sparkling of cut diamonds.

Total internal reflection enables light to be transmitted inside thin glass fibers. The light is internally reflected off the sides of the fiber, and, therefore, follows the path of the fiber. Light can actually be transmitted around corners using a glass fiber, provided that the bends in the fiber are not too sharp, so that the light always strikes the sides of the fiber at angles greater than the critical angle. The whole field of *fiber optics,* with its many useful applications, is based on this effect.

Dispersion

When a wave is refracted into a dielectric medium whose refractive index *varies* with wavelength then the angle of refraction also varies with wavelength. If the incident wave is not monochromatic, but is, instead, composed of a mixture of waves of different wavelengths, then each component wave is refracted through a *different* angle. This phenomenon is called *dispersion.*

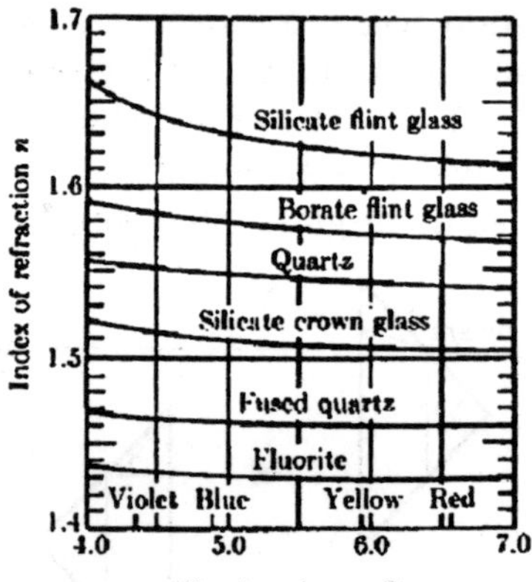

Fig. Refractive Indices of Some Common Materials as Functions of Wavelength.

The refractive indices of some common materials as functions of wavelength in the visible range. It can be seen that the refractive index always *decreases* with increasing wavelength in the visible range. In other words, violet light is always refracted *more strongly* than red light.

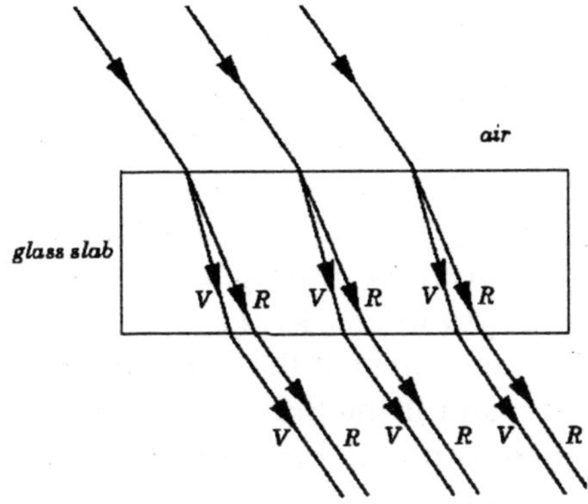

Fig. Dispersion of Light by a Parallel-Sided Glass Slab.

Suppose that a parallel-sided glass slab is placed in a beam of white light. Dispersion takes place inside the slab, but, since the rays which emerge from the slab all run *parallel* to one another, the dispersed colours recombine to form white light again, and no dispersion is observed except at the very edges of the beam. . It follows that the dispersion of white light through a parallel-sided glass slab is not generally a noticeable effect.

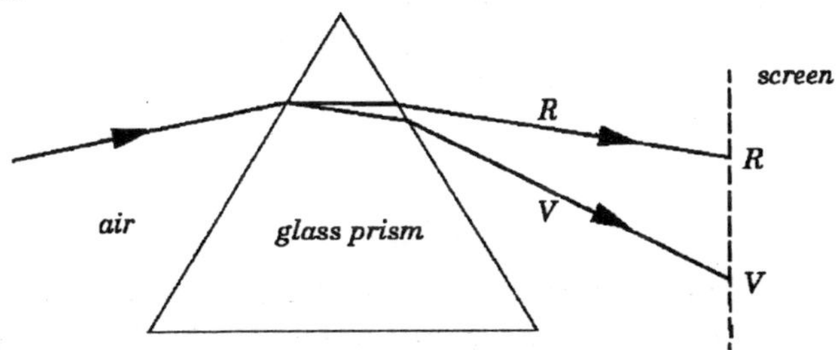

Fig. Dispersion of Light by a Glass Prism

Suppose that a glass prism is placed in a beam of white light. Dispersion takes place inside the prism, and, since the emerging rays are *not parallel* for different colours, the dispersion is clearly noticeable, especially if the emerging rays are projected onto a screen which is placed a long way from the prism. It is clear that a glass prism is far more effective at separating white light into its component colours than a parallel-sided glass slab (which explains why prisms are generally employed to perform this task).

RAINBOWS

The most well-known, naturally occurring phenomenon which involves the dispersion of light is a *rainbow*. A rainbow is an *arc* of light, with an angular radius of 42°, centred on a direction which is *opposite* to that of the Sun in the sky (*i.e.*, it is centred on the direction of propagation of the Sun's rays). Thus, if the Sun is low in the sky (*i.e.*, close to the horizon) we see almost a full semi-circle. If the Sun is higher in the sky we see a smaller arc, and if the Sun is more than 42° above the horizon then there is no rainbow (for viewers on the Earth's surface).

Observers on a hill may see parts of the rainbow below the horizontal: *i.e.*, an arc greater than a semi-circle. Passengers on an airplane can sometimes see a full circle.

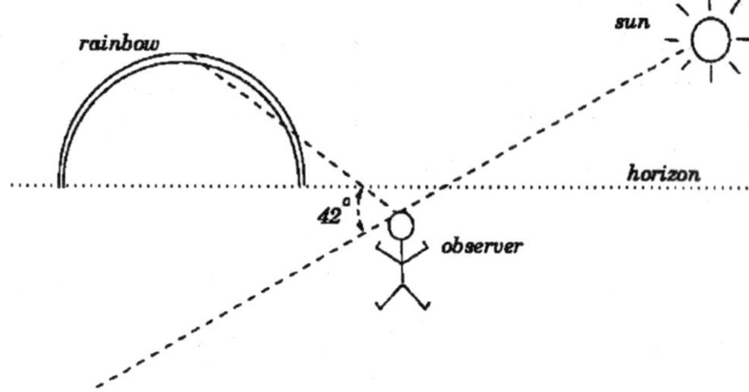

Fig. A Rainbow

The colours of a rainbow vary smoothly from red on the outside of the arc to violet on the inside. A rainbow has a diffuse inner edge, and a sharp outer edge. Sometimes a *secondary arc* is observed. This is fainter and larger (with an angular radius of 50°) than the primary arc, and the order of the colours is reversed (*i.e.*, red is on the inside, and violet on the outside).

The secondary arc has a diffuse outer edge, and a sharp inner edge. The sky between the two arcs sometimes appears to be less bright than the sky elsewhere. This region is called *Alexander's dark band*, in honour of Alexander of Aphrodisias who described it some 1800 years ago.

Rainbows have been studied since ancient times. Aristotle wrote extensively on rainbows in his *De Meteorologica,* and even speculated that a rainbow is caused by the reflection of sunlight from the drops of water in a cloud.

The first scientific study of rainbows was performed by Theodoric, professor of theology at Freiburg, in the fourteenth century. He studied the path of a light-ray through a spherical globe of water in his laboratory, and suggested that the globe be thought of as a model of a single falling raindrop. A ray, from the Sun, entering the drop, is refracted at the air-water interface, undergoes internal reflection from the inside surface of the drop, and then leaves the drop in a backward direction, after being again refracted at the surface. Thus, looking away from the Sun, towards a cloud of raindrops, one sees an enhancement of light due to these rays. Theodoric did not explain why this enhancement is concentrated at a particular angle from the direction of the Sun's rays, or why the light is split into different colours.

The first person to give a full explanation of how a rainbow is formed was René Descartes. He showed mathematically that if one traces the path through a spherical raindrop of parallel light-rays entering the drop at different points on its surface, each emerges in a different direction, but there is a concentration of emerging rays at an angle of 42° from the reverse direction to the incident rays, in exact agreement with the observed angular size of rainbows. Furthermore, since some colours are refracted more than others in a raindrop, the "rainbow angle" is slightly different for each colour, so a raindrop disperses the Sun's light into a set of nearly overlapping coloured arcs.

It shows parallel light-rays entering a spherical raindrop. Only rays entering the upper half contribute to the rainbow effect. Let us follow the rays, one by one, from the top down to the middle of the drop. We observe the following pattern. Rays which enter near the top of the drop emerge going in almost the reverse direction, but a few degrees below the horizontal. Rays entering a little further below the top emerge at a greater angle below the horizontal. Eventually, we reach a critical ray, called the *rainbow ray*, which emerges in an angle 42° below the horizontal.

Rays entering the drop lower than the rainbow ray emerge at an angle less than 42°. Thus, the rainbow ray is the one which deviates *most* from the reverse direction to the incident rays.

This variation, with 42°being the maximum angle of deviation from the reverse direction, leads to a bunching of rays at that angle, and, hence, to an unusually bright arc of reflected light centred around 42° from the reverse direction. The arc has a sharp outer edge, since reflected light *cannot* deviate by more than 42° from the reverse direction, and a diffuse inner edge, since light *can* deviate by less than 42° from the reverse direction: 42° is just the *most likely* angle of deviation.

Finally, since the rainbow angle varies slightly with wavelength (because the refractive index of water varies slightly with wavelength), the arcs corresponding to each colour appear at slightly different angles relative to the reverse direction to the incident rays.

We expect violet light to be refracted more strongly than red light in a raindrop. The red arc deviates slightly more from the reverse direction to the incident rays than the violet arc. In other words, violet is concentrated on the inside of the rainbow, and red is concentrated on the outside.

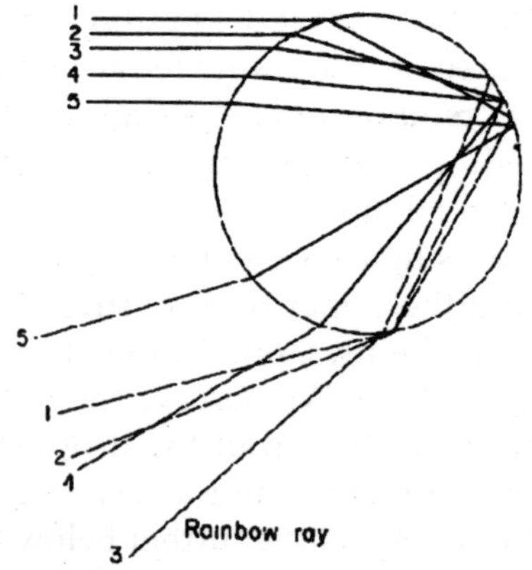

Fig. Descarte's Theory of the Rainbow.

Descartes was also able to show that light-rays which are internally reflected *twice* inside a raindrop emerge concentrated at an angle of 50° from the reverse direction to the incident rays. Of course, this angle corresponds exactly to the angular size of the secondary rainbow sometimes seen outside the first.

This rainbow is naturally less intense than the primary rainbow, since a light-ray loses some of its intensity at each reflection or refraction event. Note that 50°represents the angle of *maximum* deviation of doubly reflected light from the reverse direction (*i.e.*, doubly reflected light can deviate by more than this angle, but not by less). Thus, we expect the secondary rainbow to have a diffuse outer edge, and a sharp inner edge. We also expect doubly reflected violet light to be refracted more strongly in a raindrop than doubly reflected red light.

The red secondary arc deviates slightly less from the reverse direction to the incident rays than the violet secondary arc. In other words, red is concentrated on the inside of the secondary rainbow, and violet on the outside. Since no reflected light emerges between the primary and secondary rainbows (*i.e.*, in the angular range 42° to 50°, relative to the reverse direction), we naturally expect this region of the sky to look slightly less bright than the other surrounding regions of the sky, which explains Alexander's dark band.

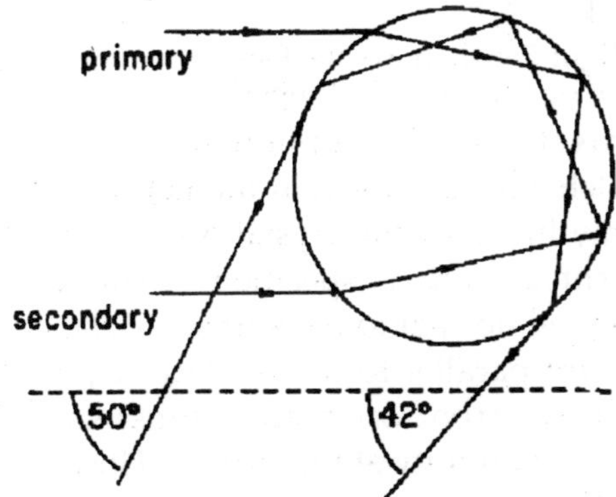

Fig. Rainbow Rays for the Primary and Ssecondary Arcs of a Rainbow.

ANISOTROPIC MEDIA

Notice that Snell's law makes the implicit assumption that rays of light move in the direction of the light's wave vector, i. e., normal to the wave fronts. This is valid only when the optical medium is isotropic, i. e., the wave frequency depends only on the magnitude of the wave vector, not on its direction.

Certain kinds of crystals, such as those made of calcite, are not isotropic — the speed of light in such crystals, and hence the wave frequency, depends on the orientation of the wave vector. As an example, the angular frequency in an isotropic medium might take the form

$$\omega = \left[\frac{c_1^2 (k_x + k_y)^2}{2} + \frac{c_2^2 (c_x - k_y)^2}{2} \right]^{1/2},$$

where c_1 is the speed of light for waves in which $k_y = k_x$, and c_2 is its speed when $k_y = k_x$.

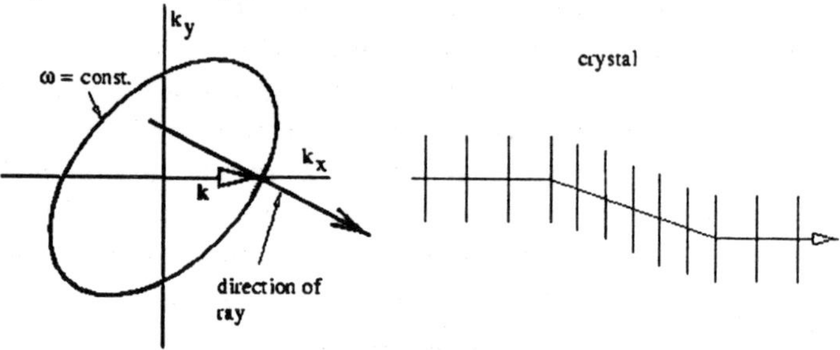

Fig. The Right Panel Shows the Fate of a Light Ray Normally Incident on the Face of a Properly Cut Calcite Crystal.

The Anisotropic Dispersion Relation Which Gives Rise to this Behavior Is Shown in the Left Panel. An example in which a ray hits a calcite crystal oriented so that constant frequency contours are as specified in equation. The wave vector is oriented normal to the surface of the crystal, so that wave fronts are parallel to this surface. Upon entering the crystal, the wave front orientation must stay the same to preserve phase continuity at the surface. However, due to the anisotropy of the dispersion relation for light in the crystal, the ray direction changes in the right panel.

This behavior is clearly inconsistent with the usual version of Snell's law!

It is possible to extend Snell's law to the anisotropic case. However, we will not present this here. The following discussions of optical instruments will always assume that isotropic optical media are used.

THIN LENS FORMULA AND OPTICAL INSTRUMENTS

Given the laws of reflection and refraction, one can see in principle how the passage of light through an optical instrument could be traced. For each of a number of initial rays, the change in the direction of the ray at each mirror surface or refractive index interface can be calculated. Between these points, the ray traces out a straight line.

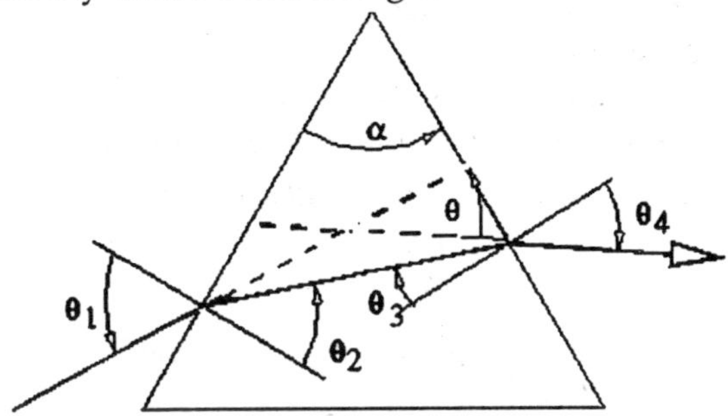

Fig. Bending of a Ray of Light as it Passes Through a Prism.

Though simple in conception, this procedure can be quite complex in practice. However, the procedure simplifies if a number of approximations, collectively called the *thin lens approximation*, are valid. We begin with the calculation of the bending of a ray of light as it passes through a prism.

The pieces of information needed to find θ, the angle through which the ray is deflected are as follows: The geometry of the triangle defined by the entry and exit points of the ray and the upper vertex of the prism leads to

$$\alpha + (\pi/2 - \theta_2) + (\pi/2 - \theta_3) = \pi,$$

which simplifies to

$$\alpha = \theta_2 + \theta_3.$$

Snell's law at the entrance and exit points of the ray tell us that

$$n = \frac{\sin(\theta_1)}{\sin(\theta_2)} \quad n = \frac{\sin(\theta_4)}{\sin(\theta_3)},$$

where n is the index of refraction of the prism. One can also infer that

$$\theta = \theta_1 + \theta_4 - \alpha.$$

This comes from the fact that the the sum of the internal angles of the shaded quadrangle is $(\pi/2 - \theta_1) + \alpha + (\pi/2 - \theta_4) + (\pi/2 - \theta_4) + (\pi + \theta) = 2\pi$.

Combining equations allows the ray deflection θ to be determined in terms of θ_1 and α, but the resulting expression is very messy. However, great simplification occurs if the following conditions are met:

- The angle $\alpha << 1$.
- The angles $\theta_1 \ll 1$ and $\theta_4 \ll 1$.

With these approximations it is easy to show that

$$\theta = \alpha(n-1) \quad \text{(small angles)}.$$

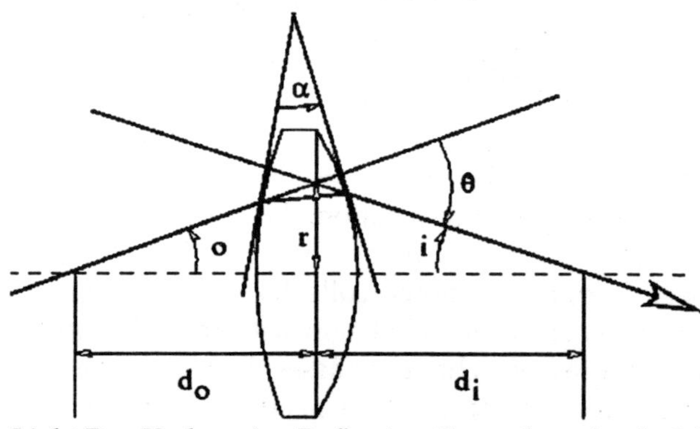

Fig. Light Ray Undergoing Deflection Through an Angle Q by a Lens. The Angle a is the Angle between the Tangents to the Entry and Exit Points of the Ray on the Lens.

Generally speaking, lenses and mirrors in optical instruments have curved rather than flat surfaces. However, we can still use the laws for reflection and refraction by plane surfaces as long as the segment of the surface on which the wave packet impinges is not curved very much on the scale of the wave packet dimensions. This condition is easy to satisfy with light impinging on ordinary optical instruments. In this case, the deflection of a ray of light is given by equation if α is defined as the intersection of the tangent lines to the entry and exit points of the ray.

A *positive lens* is thicker in the center than at the edges. The angle α between the tangent lines to the two surfaces of the lens at a distance τ from the central axis takes the form $\alpha = C_{\tau}$, where C is a constant. The deflection angle of a beam hitting the lens a distance τ from the center is therefore $\theta = C\tau$ $(n - 1)$. The angles o and i sum to the deflection angle: $o + i = \theta$ $= C\tau\ (n - 1)$. However, to the extent that the small angle approximation holds, $o = \tau/d_o$ and $i = \tau/d_i$ where d_o is the distance to the object and d_i is the distance to the image of the object. Putting these equations together and cancelling the τ results in the *thin lens formula*:

$$\frac{1}{d_o} + \frac{1}{d_i} = C(n-1) \equiv \frac{1}{f}.$$

The quantity f is called the *focal length* of the lens. Notice that $f = d_i$ if the object is very far from the lens, i. e., if d_o is extremely large.

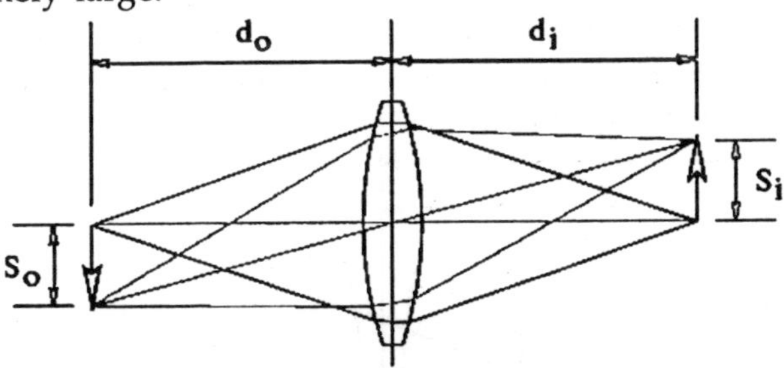

Fig. A Positive Lens Producing an Image on the Right of the Arrow on the Left.

The image is produced by all of the light from each point on the object falling on a corresponding point in the image. If the arrow on the left is an illuminated object, an *image* of the arrow will appear at right if the light coming from the lens is allowed to fall on a piece of paper or a ground glass screen.

The size of the object S_o and the size of the image S_i are related by simple geometry to the distances of the object and the image from the lens:

$$\frac{S_i}{S_o} = \frac{d_i}{d_o}.$$

Notice that a positive lens inverts the image.

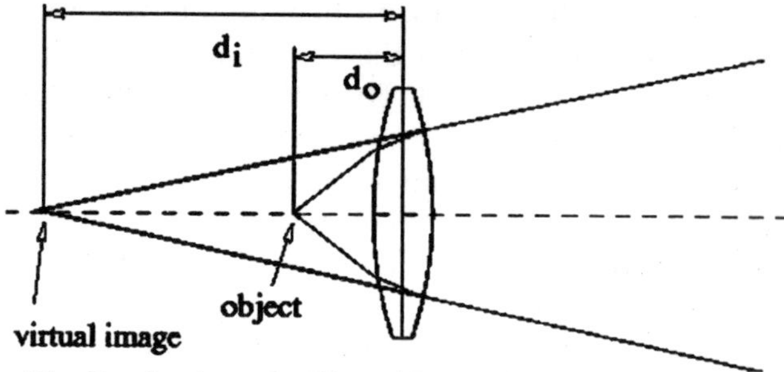

Fig. Production of a Virtual Image by a Positive Lens.

An image will be produced to the right of the lens only if $d_o > f$. If $d_o < f$, the lens is unable to converge the rays from the image to a point. However, in this case the backward extension of the rays converge at a point called a *virtual image,* which in the case of a positive lens is always farther away from the lens than the object. The thin lens formula still applies if the distance from the lens to the image is taken to be negative. The image is called virtual because it does not appear on a ground glass screen placed at this point.

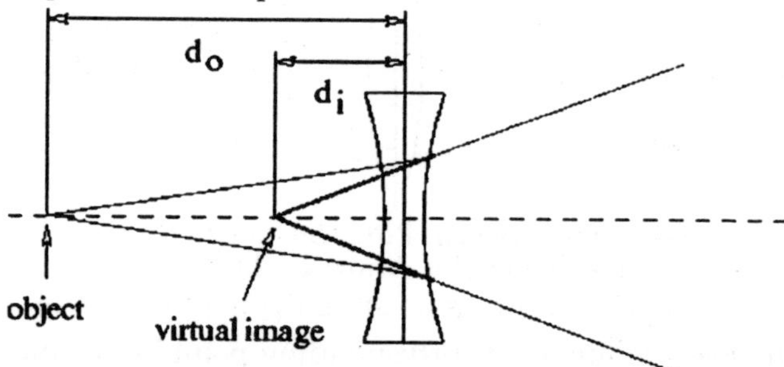

Fig. Production of a Virtual Image by a Negative Lens.

A *negative lens* is thinner in the center than at the edges and produces only virtual images. The virtual image produced by a negative lens is closer to the lens than is the object. Again, the thin lens formula is still valid, but both the distance from the image to the lens and the focal length must be taken as negative. Only the distance to the object remains positive.

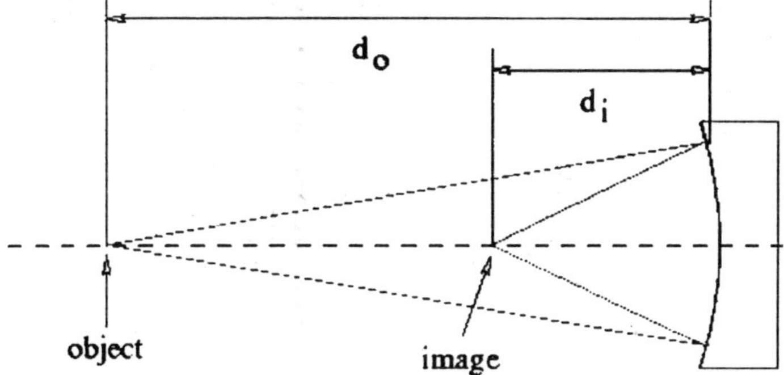

Fig. Production of a Real Image by a Concave Mirror.

Curved mirrors also produce images in a manner similar to a lens. A concave mirror works in analogy to a positive lens, producing a real or a virtual image depending on whether the object is farther from or closer to the mirror than the mirror's focal length.

A convex mirror acts like a negative lens, always producing a virtual image. The thin lens formula works in both cases as long as the angles are small.

FERMAT'S PRINCIPLE

An alternate approach to geometrical optics can be developed from Fermat's principle. This principle states that light waves of a given frequency traverse the path between two points which takes the least time. The most obvious example of this is the passage of light through a homogeneous medium in which the speed of light doesn't change with position.

In this case shortest time is equivalent to the shortest distance between the points, which, as we all know, is a straight line. Thus, Fermat's principle is consistent with light traveling in a straight line in a homogeneous medium.

Fermat's principle can also be used to derive the laws of reflection and refraction.

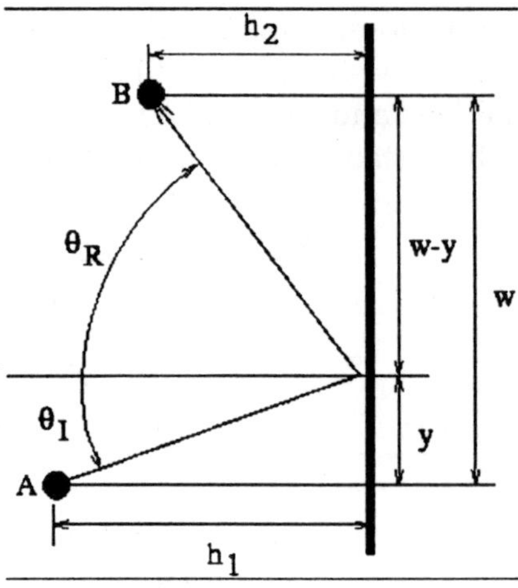

Fig. Definition Sketch for Deriving the Law of Reflection from Fermat's Principle. θ_I is the Angle of Incidence and θ_R the Angle of Reflection.

A candidate ray for reflection in which the angles of incidence and reflection are not equal. The time required for the light to go from point A to point B is

$$t = \left[\left\{ h_1^2 + y^2 \right\}^{1/2} + \left\{ h_2^2 + (w - y)^2 \right\}^{1/2} \right] / c$$

where c is the speed of light. We find the minimum time by differentiating t with respect to y and setting the result to zero, with the result that

$$\frac{y}{\left\{ h_1^2 + y^2 \right\}^{1/2}} = \frac{w - y}{\left\{ h_2^2 + (w - y)^2 \right\}^{1/2}}.$$

However, we note that the left side of this equation is simply $\sin \theta_I$, while the right side is $\sin \theta_R$, so that the minimum time condition reduces to $\sin \theta_I = \sin \theta_R$ or $\theta_I = \theta_R$, which is the law of reflection.

A similar analysis may be done to derive Snell's law of refraction. The speed of light in a medium with refractive index n is c/n, where is its speed in a vacuum. Thus, the time required for light to go some distance in such a medium is times the time light takes to go the same distance in a vacuum the time required for light to go from A to B becomes

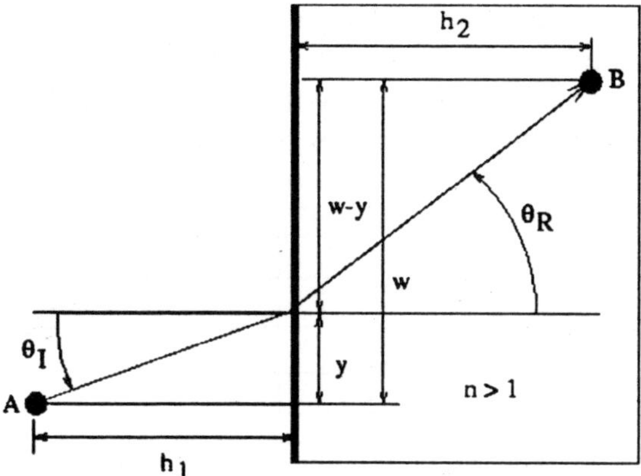

Fig. Definition Sketch for Deriving Snell's Law of Refraction from Fermat's Principle. The Shaded Area has Index of Refraction $n > 1$.

$$t = \left[\left\{h_1^2 + y^2\right\}^{1/2} + n\left\{h_2^2 + (w - y)^2\right\}^{1/2}\right]/c.$$

This results in the condition

$$\sin\theta_I = n\sin\theta_R$$

where θ_R is now the refracted angle. We recognize this result as Snell's law.

The minimum time may actually be a *local* rather than a global minimum the global minimum distance from A to B is still just a straight line between the two points! In fact, light starting from point A will reach point B by *both* routes — the direct route and the reflected route.

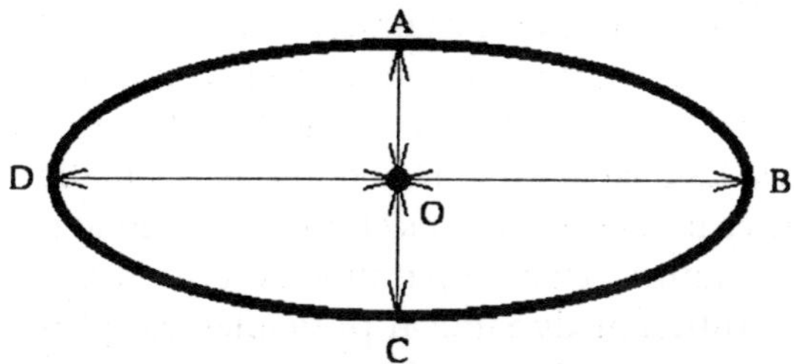

Fig. Ellipsoidal Mirror Showing Minimum and Maximum Time Rays from the Center of the Ellipsoid to the Mirror Surface and Back again.

It turns out that trajectories allowed by Fermat's principle don't strictly have to be minimum time trajectories. They can also be *maximum* time trajectories In this case light emitted at point O can be reflected back to point O from four points on the mirror, A, B, C, and D. The trajectories O-A-O and O-C-O are minimum time trajectories while O-B-O and O-D-O are maximum time trajectories.

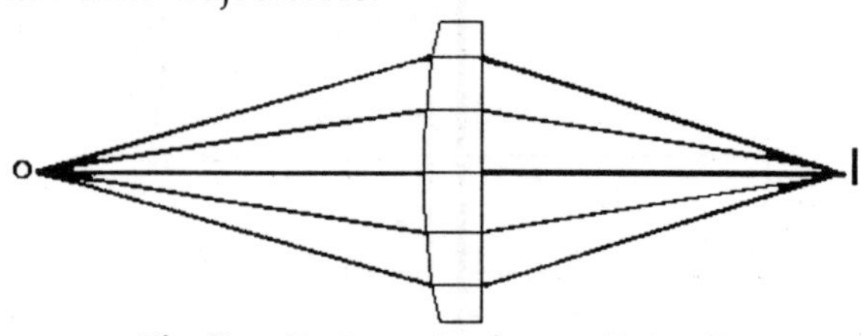

Fig. Ray Trajectories from a Point O being Focused to Another Point I by a Lens.

All the rays from point O which intercept the lens end up at point I. This would seem to contradict Fermat's principle, in that only the minimum (or maximum) time trajectories should occur. However, a calculation trajectories in this particular case take the same time. Thus, the light cannot choose one trajectory over another using Fermat's principle and all of the trajectories are equally favored. Note that this inference applies not to just any set of trajectories, but only those going from one focal point to another.

GEOMETRICAL OPTICS AND NEWTON'S LAWS

The question that motivates us to study physics is "What makes things go?" The answers we conceive to this question constitute the subject of dynamics. This is in contrast to the question we have primarily addressed so far, namely "How do things go?" As noted earlier, the latter question is about kinematics. Extensive preparation in the kinematics of waves and particles in relativistic spacetime is needed to intelligently address dynamics. This preparation is now complete.

Three different dynamical principles based respectively on pre-Newtonian, Newtonian, and quantum mechanical thinking. The last two are still in common use, and we first show that they are consistent with each other in the realm in which Newtonian and quantum mechanics overlap, i. e., in the geometrical optics limit of quantum mechanics. For simplicity, this relationship is first developed in one dimension in the non-relativistic limit. Higher dimensions require the introduction of partial derivatives, and the relativistic case will be considered later.

Only forces which are "conservative" in Newtonian mechanics, i. e., exhibit a potential energy, have a correspondence in the quantum mechanical world. Forces of this type receive our primary attention. Certain ancillary concepts in Newtonian mechanics such as work and power are introduced at this stage.

FUNDAMENTAL PRINCIPLES OF DYNAMICS

Newtonian Dynamics

In contrast to earthly behavior, the motions of celestial objects seem effortless. No obvious forces act to keep the planets in motion around the sun. In fact, it appears that celestial objects simply coast along at constant velocity unless something acts on them. The Newtonian view of dynamics — objects change their velocity rather than their position when a force is exerted on them — is expressed by Newton's second law:

$$F = ma \text{ (newton's second law)}$$

where F is the force exerted on a body, m is its mass, and a is its acceleration. Newton's first law, which states that an object remains at rest or in uniform motion unless a force acts on it, is actually a special case of Newton's second law which applies when $F = 0$.

It is no wonder that the first successes of Newtonian mechanics were in the celestial realm, namely in the predictions of planetary orbits. It took Newton's genius to realize that the same principles which guided the planets also applied to the earthly realm as well. In the Newtonian view, the tendency of objects to stop when we stop pushing on them is simply a consequence of frictional forces opposing the motion. Friction, which is so important on the earth, is negligible for planetary motions, which is why Newtonian dynamics is more obviously valid for celestial bodies.

Note that the principle of relativity is closely related to Newtonian physics and is incompatible with pre-Newtonian views. After all, two reference frames moving relative to each other cannot be equivalent in the pre-Newtonian view, because objects with nothing pushing on them can only come to rest in one of the two reference frames! Einstein's relativity is often viewed as a repudiation of Newton, but this is far from the truth — Newtonian physics makes the theory of relativity possible through its invention of the principle of relativity. Compared with the differences between pre-Newtonian and Newtonian dynamics, the changes needed to go from Newtonian to Einsteinian physics constitute minor tinkering.

Quantum Dynamics

In quantum mechanics, particles are represented by matter waves, with the absolute square of the wave displacement yielding the probability of finding the particle. The behavior of particles thus follows from the reflection, refraction, diffraction, and interference of the associated waves. The connection with Newtonian dynamics comes from tracing the trajectories of matter wave packets. Changes in the speed and direction of motion of these packets correspond to the accelerations of classical mechanics. When wavelengths are small compared to the natural length scale of the problem at hand, the wave packets can be made small, thus pinpointing the position of the associated particle, without generating excessive uncertainty in the particle's momentum. This is the geometrical optics limit of quantum mechanics.

ONE-DIMENSIONAL NON-RELATIVISTIC CASE

Louis de Broglie[9.1] made an analogy between matter waves and light waves, pointing out that wave packets of light change their velocity as the result of spatial variations in the index of refraction of the medium in which they are travelling. This behavior comes about because the dispersion relation for light traveling through a medium with index of refraction n is $\omega = kc/n$, so that the group velocity, $u_s = d\omega/dk = c/n$. Thus, when n increases, u_s decreases, and vice versa.[9.2]

The problem for matter waves is to determine how analogous modifications can be made to the free particle dispersion relation so as to produce accelerations of wave packets consistent with Newtonian dynamics in the geometrical optics limit.

A simple guess as to how to modify the free particle dispersion relation for matter waves, limiting consideration initially to the one-dimensional, non-relativistic case. As in many situations in physics, the simple guess turns out to be correct! This leads to a connection between wave dynamics and Newtonian mechanics.

The dispersion relation for free matter waves is $\omega = (k^2c^2 + \mu^2)^{1/2}$. In the non-relativistic limit $k^2 c^2 << \mu^2$ and this equation can be approximated as

$$\omega = \mu(1 + k^2c^2/\mu^2)^{1/2} \approx \mu + k^2c^2/(2\mu).$$

Let us modify this equation by adding a term S' which depends on x, and which plays a role analogous to the index of refraction for light:

$$\omega = S'(x) + \mu + k^2c^2/(2\mu) = S(x) + k^2c^2/(2\mu).$$

The rest frequency has been made to disappear on the right side of the above equation by defining $S = S' + \mu$. This is done to simplify the notation.

Let us now imagine that all parts of the wave governed by this dispersion relation oscillate in phase. The only way this can happen is if ω is constant, i. e., it takes on the same value in all parts of the wave. It turns out we can do this if S is not a function of time.

If ω is constant, the only way S can vary with x in equation is if the wavenumber varies in a compensating way. Thus, constant frequency and spatially varying S together imply that $k = k(x)$. Solving equation for k yields

$$k(x) = \left[\frac{2\mu[\omega - S(x)]}{c^2}\right]^{1/2}.$$

Since ω is constant, the wavenumber becomes smaller and the wavelength larger as the wave moves into a region of increased S.

In the geometrical optics limit, we assume that S doesn't change much over one wavelength so that the wave remains reasonably sinusoidal in shape with approximately constant wavenumber over a few wavelengths. However, over distances of many wavelengths the wavenumber and amplitude of the wave are allowed to vary considerably.

As yet we have no idea what causes S or where it comes from. For now we simply explore the consequences of its presence, especially in the geometric optics limit in which quantum dynamics gives way to Newtonian dynamics.

The group velocity calculated from the dispersion relation given by equation is

$$u_g = \frac{d\omega}{dk} = \frac{kc^2}{\mu} = \left(\frac{2c^2(\omega - S)}{\mu}\right)^{1/2}$$

where k is eliminated in the last step with the help of equation The resulting equation tells us how the group velocity varies as a matter wave traverses a region of slowly varying S. Thus, as S increases, u_g decreases and vice versa.

We can now calculate the acceleration of a wave packet resulting from the spatial variation in S. We assume that $x(t)$ represents the position of the wave packet, so that $u_g = dx/dt$. Using the chain rule $du_g/dt = (du_g/dx)(dx/dt) = (du_g/dx)u_g$, we find

$$a = \frac{du_g}{dt} = \frac{du_g}{dx}u_g = \frac{du_g^2/2}{dx} = -\frac{c^2}{\mu}\frac{dS}{dx} = -\frac{\hbar}{m}\frac{dS}{dx}.$$

The group velocity is eliminated in favor of S by squaring equation and substituting the result into equation.

Using Newton's second law to infer the force from the acceleration, we find

$$F = ma = -\frac{dU}{dx} \text{(force from potential energy)},$$

where we have defined $U \equiv \hbar S$. The quantity U is called the potential energy. We have thus established a relationship between quantum mechanical and Newtonian dynamics, in that $U = \hbar S$ dictates the form of the force in Newton's second law, while S governs the refraction of matter waves. A force which equals minus the derivative of some potential energy $U(x)$ is called conservative. Certain forces such as friction are not conservative. At present we will deal mainly with conservative forces.

POTENTIAL ENERGY

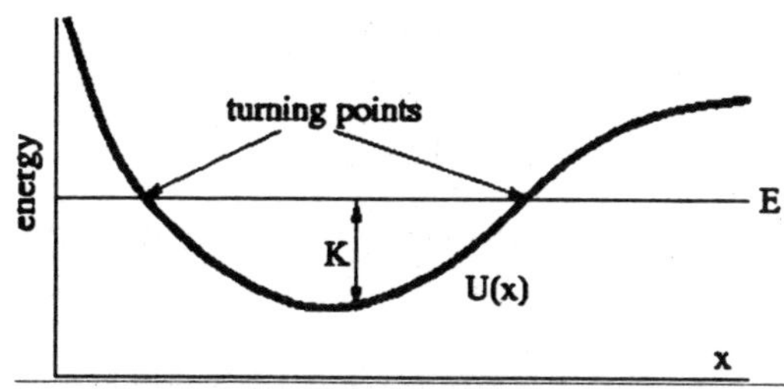

Fig. Example of Spatially Variable Potential Energy $U(x)$ with Fixed Total Energy E. The Kinetic Energy $k = E - U$ is Zero where the E and U Lines Cross. These points are Called Turning Points.

Recall that the momentum is related to the group velocity by

$$\Pi = mu_g \text{ (momentum)}$$

in the nonrelativistic case. Multiplying equation by $\hbar$ and eliminating k and μ in favor of Π and m results in

$$E = U + \Pi^2/(2m) = U + K \text{ (total energy)}$$

where we have used $U = \hbar S$ and $\hbar m = mc^2$, and where

$$K \equiv \Pi^2/(2m) = mu_g^2/2 \text{ (kinetic energy)}$$

is the nonrelativistic kinetic energy.

The energy E, which we now refer to for clarity as the total energy, is constant since it is proportional to the (constant) frequency ω. We say that the total energy is conserved. Thus, increases in the potential energy coincide with decreases in the kinetic energy and vice versa. In Newtonian dynamics the kinetic energy cannot be negative, since it is the product of half the mass and the square of the velocity, both of which are positive. Thus, a particle with total energy E and potential energy $U$is forbidden to venture into regions in which the kinetic energy $K = E - U$ is negative.

The points at which the kinetic energy is zero are called turning points. This is because a particle decreases in speed as it approaches a turning point, stops there for an instant, and reverses direction.

Note also that a particle with a given total energy always has the same speed at some point x regardless of whether it approaches this point from the left or the right:

$$\text{speed} = |u_g| = |\pm[2(E-U)/m]^{1/2}|.$$

Gravity as a Conservative Force

An example of a conservative force is gravity. An object of mass m near the surface of the earth has the gravitational potential energy

$$U = mgz \text{ (gravity near earth's surface)}$$

where z is the height of the object and $g = 9.8 \text{ ms}^{-2}$ is the local value of the gravitational field near the earth's surface. Notice that the gravitational potential energy increases upward. The speed of the object in this case is $|u_g| \backslash [2(E - mgz)/m]^{1/2}$. If $|u_g|$ is known to equal the constant value u_0 at elevation $z = 0$, then equations tell us that $u_0 = (2E/m)^{1/2}$ and $|u_g| = (u_0^2 - 2gz)^{1/2}$.

There are certain types of questions which energy conservation cannot directly answer. For instance if an object is released at elevation h with zero velocity at $t = 0$, at what time will it reach $z = 0$ under the influence of gravity? In such cases it is often easiest to return to Newton's second law. Since the force on the object is $F = -dU/dz = -mg$ in this case, we find that the acceleration is $a = F/m = -mg/m = -g$. However, $a = du/dt = d^2z/dt^2$, so

$$u = -gt + C_1 \quad z = -gt^2/2 + C_1t + C_2 \text{ (constnat gravity)},$$

where C_1 and C_2 are constants to be determined by the initial conditions.

These results can be verified by differentiating to see if the original acceleration is recovered. Since $u = 0$ and $z = h$ at $t = 0$, we have $C_1 = 0$ and $C_2 = h$. With these results it is easy to show that the object reaches $z = 0$ when $t = (2h/g)^{1/2}$.

WORK AND POWER

When a force is exerted on an object, energy is transferred to the object. The amount of energy transferred is called the work done on the object. However, energy is only transferred if the object moves. The work W done is

$$W = F\Delta x$$

where the distance moved by the object is Δx and the force exerted on it is F. Notice that work can either be positive or negative. The work is positive if the object being acted upon moves in the same direction as the force, with negative work occurring if the object moves opposite to the force.

Equation assumes that the force remains constant over the full displacement Δx. If it is not, then it is necessary to break up the displacement into a number of smaller displacements, over each of which the force can be assumed to be constant. The total work is then the sum of the works associated with each small displacement.

If more than one force acts on an object, the works due to the different forces each add or subtract energy, depending on whether they are positive or negative. The total work is the sum of these individual works.

There are two special cases in which the work done on an object is related to other quantities. If F is the total force acting on the object, then $W = F\Delta x = ma\Delta x$by Newton's second law. However, $a = dv/dt$ where v is the velocity of the object, and $\Delta x = (\Delta x/\Delta t)\Delta t \approx v\Delta t$, where Δt is the time required by the object to move through distance Δx. The approximation becomes exact when Δx and Δt become very small. Putting all of this together results in

$$W_{total} = m\frac{dv}{dt}v\Delta t = \frac{d}{dt}\left(\frac{mv^2}{2}\right)\Delta t = \Delta K \text{ (total work)},$$

where K is the kinetic energy of the object. Thus, when F is the only force, $W = W_{total}$ is the total work on the object, and this equals the change in kinetic energy of the object. This is called the work-energy theorem, and it demonstrates that work really is a transfer of energy to an object.

The other special case occurs when the force is conservative, but is not necessarily the total force acting on the object. In this case

$$W_{cons} = -\frac{dU}{dx}\Delta x = -\Delta U \text{ (conservative force)},$$

where ΔU is the change in the potential energy of the object associated with the force of interest.

The power associated with a force is simply the amount of work done by the force divided by the time interval Δt over which it is done. It is therefore the energy per unit time transferred to the object by the force of interest. From equation we see that the power is

$$P = \frac{F\Delta x}{\Delta t} = Fv \text{ (Power)},$$

where v is the velocity at which the object is moving. The total power is just the sum of the powers associated with each force. It equals the time rate of change of kinetic energy of the object:

$$P_{total} = \frac{W_{total}}{\Delta t} = \frac{dK}{dt} \text{ (total power)}.$$

PARTIAL DERIVATIVES

In order to understand the generalization of Newtonian mechanics to two and three dimensions, we first need to understand a new type of derivative called the partial derivative.

To motivate the discussion, let us use the chain rule to take the ordinary derivative of $f(x, y) = x^2 + y^2$ where y is some function of x, say $y = x^2 + x$:

$$\frac{df}{dx} + 2x + 2y\frac{dy}{dx} = 2x + 2y.(3x^2 + 1).$$

(At this point we don't really care what form $y = y(x)$ takes.) At times we may wish to take the derivative of f with respect to x while ignoring the possible dependence of y on x. A derivative of this type has special notation — the "d" is replaced by "∂" and the derivative is called a partial derivative. Thus, the partial derivative of f with respect to x is

$$\frac{\partial f}{\partial x} = 2x.$$

Notice that the partial derivative is actually simpler to evaluate than an ordinary derivative, because only the explicit dependence of the function on the differentiating variable need be considered — all other variables are taken to be constant.

MOTION IN TWO AND THREE DIMENSIONS

When a matter wave moves through a region of variable potential energy in one dimension, only the wavenumber changes. In two or three dimensions the wave vector can change in both direction and magnitude. This complicates the calculation of particle movement. However, we already have an example of how to handle this situation, namely, the refraction of light. In that case Snell's law tells us how the direction of the wave vector changes, while the dispersion relation combined with the constancy of the frequency gives us information about the change in the magnitude of the wave vector. For matter waves a similar procedure works, though the details are different, because we seek the consequences of a change in potential energy rather than a change in the index of refraction.

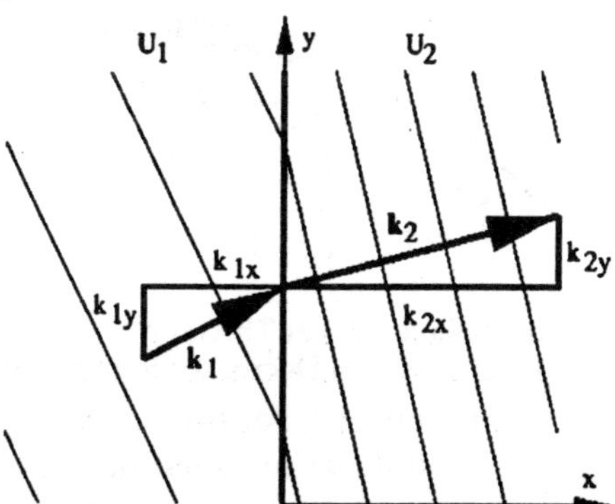

Fig. Refraction of a Matter wave by a Discontinuity in Potential Energy. The Component of the Wave Vector Parallel to the Discontinuity, k_y, Doesn't Change, so $k_{1y} = k_{2y}$.

The refraction of matter waves at a discontinuity in the potential energy. Let us suppose that the discontinuity occurs at $x = 0$. If the matter wave to the left of the discontinuity is $\psi_1 = \sin(k_{1x}x + k_{1y}y - \omega_1 t)$ and to the right is $\psi_2 = \sin(k_{2x}x + k_{2y}y - \omega_2 t)$, then the wavefronts of the waves will match across the discontinuity for all time only if $\omega_1 = \omega_2 \equiv \omega$ and $k_{1y} = k_{2y} \equiv k_y$. We are already familiar with the first condition from the one-dimensional problem, so the only new ingredient is the constancy of the y component of the wave vector.

In two dimensions the momentum is a vector: $\Pi = mu$ when $|u| \ll c$, where u is the particle velocity. Furthermore, the kinetic energy is $K = m\,|u|^2/2 = |\Pi|^2/(2m)\ (\Pi_x^2 + \Pi_y^2)/(2m)$. The relationship between kinetic, potential, and total energy is unchanged from the one-dimensional case, so we have

$$E = U + (\Pi_x^2 + \Pi_y^2)/(2m) = \text{constant}.$$

The de Broglie relationship tells us that $\Pi = \hbar k$, so the constancy of k_y across the discontinuity in U tells us that

$$\Pi_y = \text{constant}$$

there.

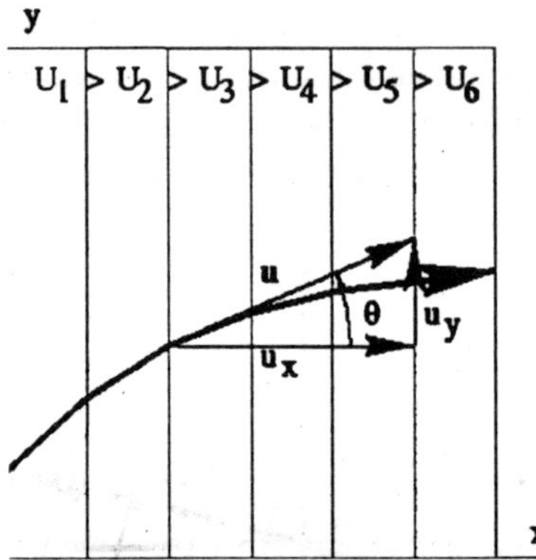

Fig. Trajectory of a Wave Packet Through a Variable Potential Energy, $U(x)$, which Decreases to the Right.

Let us now approximate a continuously variable $U(x)$ by a series of steps of constant U oriented normal to the x axis. The above analysis can be applied at the jumps or discontinuities in U between steps with the result that equations are valid across all discontinuities. If we now let the step width go to zero, these equations then become valid for Ucontinuously variable in x.

An example from classical mechanics of a problem of this type is a ball rolling down an inclined ramp with an initial velocity component across the ramp. The potential energy decreases in the down ramp direction, resulting in a force down the ramp. This accelerates the ball in that direction, but leaves the component of momentum across the ramp unchanged.

Fig. Classical Mechanics

Using the procedure which we invoked before, we find the force components associated with $U(x)$ in the x and y directions to be $F_x = -dU/dx$ and $F_y = 0$. This generalizes to

$$F = -\left(\frac{\partial U}{\partial x}, \frac{\partial U}{\partial y}, \frac{\partial U}{\partial z}\right) \text{(3-D conservative force)}$$

in the three-dimensional case where the orientation of constant U surfaces is arbitrary. It is also valid when $U(x, y, z)$ is not limited to a simple ramp form, but takes on a completely arbitrary structure.

The definitions of work and power are slightly different in two and three dimensions. In particular, work is defined

$$W = F.\Delta x$$

where Δx is now a vector displacement. The vector character of this expression yields an additional possibility over the one dimensional case, where the work is either positive or negative depending on the direction of Δx relative to F. If the force and the displacement of the object on which the force is acting are perpendicular to each other, the work done by the force is actually zero, even though the force and the displacement both have non-zero magnitudes. The power exhibits a similar change:

$$P = \text{F.u}$$

Thus the power is zero if an object's velocity is normal to the force being exerted on it.

Energy conservation by itself is somewhat less useful for solving problems in two and three dimensions than it is in one dimension. This is because knowing the kinetic energy at some point tells us only the magnitude of the velocity, not its direction. If conservation of energy fails to give us the information we need, then we must revert to Newton's second law, as we did in the one-dimensional case. For instance, if an object of mass m has initial velocity $u_0 = (u_0, 0)$ at location $(x, z) = (0, h)$ and has the gravitational potential energy $U = mgz$, then the force on the object is $F = (0, -mg)$. The acceleration is therefore $a = F/m = (0, -g)$. Since $a = du/dt = d^2x/dt^2$ where $x = (x, z)$ is the object's position, we find that

$$u = (C_1, -gt + C_2)\ x = (C_1 t + C_3, -gt^2/2 + C_2 t + C_4),$$

where C_1, C_2, C_3, and C_4 are constants to be evaluated so that the solution reduces to the initial conditions at $t = 0$. The specified initial conditions tell us that $C_1 = u_0$, $C_2 = 0$, $C_3 = 0$, and $C_4 = h$ in this case. From these results we can infer the position and velocity of the object at any time.

KINETIC AND POTENTIAL MOMENTUM

If you have previously taken a physics course then you have probably noticed that a rather odd symbol is used for momentum, namely Π, rather than the more commonly employed P. The reason for this peculiar usage is that there are actually two kinds of momentum, kinetic momentum and total momentum, just as there are two kinds of energy, kinetic and total.[9.3] The symbol Π represents total momentum while P represents kinetic momentum. In non-relativistic mechanics we don't need to distinguish between the two quantities, as they are generally equal to each other.

However, we will find later in the course that it is crucial to make this distinction in the relativistic case. As a general rule, the total momentum is related to a particle's wave vector via the de Broglie relation, $\Pi = \hbar k$, while the kinetic momentum is related to a particle's velocity, $P = mu/(1 - u^2/c^2)^{1/2}$.

Chapter 3

The Optical Field

Optical fields lie on the electromagnetic spectrum in the frequency range from 10^{14} to 10^{16} Hertz. The hard ultraviolet and x-ray regimes lie at frequencies above the optical region and the mid and far infrared and microwave regimes lies at frequencies below. The most distinguishing aspect of optical fields is that they are visible to the human eye. As there is little fundamental difference between physical phenomena observed for the visual spectrum and phenomena observed in the near infrared and ultraviolet, however, these regions are often included in the optical regime.

Two characteristics distinguish optics from other electromagnetic technologies. First, optical wavelengths (0.1 to 10 microns) are small compared to typical aperture sizes and measurement distances, meaning that optical fields are usually considered in far-field and weakly diffracting limits. Second, optical fields are generated and detected by atomic-scale quantum interactions rather than by antennas or resonators. The first characteristic leads one to analyse optical propagation under approximations to the Maxwell equations appropriate to the short wavelength regime. In many cases, one can neglect diffraction and consider optical excitations to consist of straight "beams." In such cases, one uses geometrical analysis based on ray tracing to understand the general behaviour of optical systems.

The second characteristic means that optical field generation and detection is not amenable to direct artificial control and that optical fields are susceptible to different and more profound sources of noise and uncertainty than other regions of the spectrum.

Optical fields are central to three technologies: sensing, communication and illumination. Sensing is the oldest of these, since cavemen had eyes, but fire for illumination came not much later and fire for smoke signal communication soon followed. In the modern world, optical sensing spans a range from microscopy to astronomical telescopy and includes still, cinematic and video photography, lidar, reflectometry, and a number of other applications. Optical sensing is increasingly dominated by electronic sensors and digital processing. Despite the long history of smoke signals, lighthouses, lanterns and signal lights, optical communications has bloomed as an industry only in the past 20 years.

The key to this development has been the explosion in demand for data transmission, the development of astonishingly low loss optical fiber and the development of semiconductor laser sources and detectors. Illumination has a longer history as a major industry than either sensing or communications. Illumination today includes the range from flashlights to laser tools for surgery. In view of the greater potential for research and engineering at the higher tech end of the scale, modern engineering courses tend to emphasize lasers over lightbulbs. There are other optical technologies which are difficult to classify in these three areas, for example, optical data storage mixes sensing, communications and illumination.

Just as there are three major optical applications, there are three main questions to be resolved in analyzing optical systems: how is the field generated?, how does it propagate? and how is it detected? These questions play a role in all three technologies, the significance and focus varies. Communications system development focuses primarily on modulation and detection technologies. Illumination system development focuses primarily on source issues. Sensor systems are most concerned with how the field propagates from the object or system being sensed to the detection system. While the nature of the illuminating field (e.g. its polarization, spectral and coherence properties) and the fidelity of the detection system are important, the enabling issue for sensing is how to detect information which can be back propagated to identify the object.

The primary reason for the emphasis on field propagation in sensing systems is that the most sensors detect parallel spatial information.

In contrast, the spatial pattern generated by most illumination systems contain little information and communication systems limit the field to a single spatial mode or a very small number of modes in a guiding fiber. Since the spatial pattern of the field is information rich in sensing applications, it is important to understand how this information is transformed by the spatial distortions of propagation.

The propagation of the optical field is described by Maxwell's equations. The optical field itself is not directly detectable, statistical correlations and intensity measures must supplement Maxwell's equations to fully understand propagation.

MAXWELL'S EQUATIONS

Maxwell's equations in differential form are

$$\nabla \times E = -\frac{\partial}{\partial t} B$$

$$\nabla \times H = J - \frac{\partial}{\partial t} D$$

$$\nabla . D = \rho$$

$$\nabla \mathrm{B} = 0$$

is the electric field, is the electric displacement, B is the magnetic induction, H is the magnetic field, J is the current density and is the charge density. The first equation is Faraday's Law and expresses the experimental fact that a time varying magnetic field induces an electromotive force. The second equation is Ampere's Circuital Law and expresses the experimental fact that a current or time varying electric field induces a magnetic field. The last two equations are Gauss's Law and express the experimental facts of Coloumb's law and charge conservation.

The field variables are related by the material equations

$$D = \varepsilon_0 E + P$$

$$B = \mu_0 H + M$$

where P is the *polarization* of material and M is the *magnetization*. In most optical materials, $M = 0$ and P is a function of E. The simplest and most common case is

$$P = \varepsilon_0 \chi_e E$$

Then

$$D = \in E$$

where

$$\in = \in_0 \varepsilon_\lambda$$

and

$$\varepsilon\lambda = 1 + \chi_e.$$

As a result of the high frequencies and small wavelengths of optical fields, charge interactions with the optical field involve quantum mechanical effects which cannot accurately be analysed by continuous models.

In optical diffraction problems, is always neglected. In principle, J is also neglected, although the current density can be used formally to add a loss component to the pemitivitty,. We also note that the vectors E and D need not be colinear, meaning that may in general be tensor valued. A medium in which is (not) tensor valued is called (isotropic) anisotropic.

THE WAVE EQUATION

Using the material relations, we substitute in the Maxwell equations to find the wave equations in isotropic media.

$$\nabla \times \nabla \times E - \mu_0 E \frac{\partial^2}{\partial t^2} E$$

$$\nabla \times \nabla \times E = \mu_0 E \frac{\partial^2}{\partial t^2} E$$

The equations are reduced to a simpler form by the vector identity

$$\nabla \times \nabla \times A = \nabla(\nabla.A) - \nabla^2 A$$

From the Gauss's Law we know that

$$\nabla.(\varepsilon E) = E.\nabla\varepsilon + \varepsilon\, \nabla.E = 0,$$

where we have assumed for the moment ε that is scalar valued. Thus,

$$\nabla.E = -E.\nabla \log(\varepsilon)$$

and

$$\nabla^2 E + \nabla(E.\nabla \log(\varepsilon)) - \mu_0 E \frac{\partial^2}{\partial t^2} E = 0$$

A medium in which $\nabla\varepsilon = 0$ (the permittivity is spatially constant) is *homogeneous*. Most common materials are approximately homogeneous. Media in which this is not the case include optical fiber, graded index lenses, and volume holograms.

In homogeneous isotropic media, the wave equations are

$$\nabla^2 E - \mu_0 E \frac{\partial^2}{\partial t^2} E = 0$$

$$\nabla^2 H - \mu_0 E \frac{\partial^2}{\partial t^2} H = 0$$

SOLUTIONS TO THE WAVE EQUATION

Fields satisfying the wave equations are 4D disturbances in space-time. Since a wave disturbance propagates without dissipation, there are no spatially bounded solutions to the wave equation in source and absorber free space. Pulse packets in localized in space-time are potential solutions, but such solutions diffractively dissipate on propagation. There are no simple exact solutions of finite energy.

Plane waves are the simplest solutions to the wave equation. Plane waves are unbounded and constant in two dimensions and propagate along a third dimension. Assuming that the constant dimensions are x and y, the derivative of the fields vanish along these dimension and the wave equation for the electric field reduces to

$$\frac{\partial^2}{\partial z^2}E_x - \mu_0 E\frac{\partial^2}{\partial t^2}E_x = 0$$

where we have chosen to write the equation only for one polarization component of the electric field. Note that for the plane wave to satisfy Gauss's law the z component of the field must be zero.

Eqn. has many possible solutions, the simplest being

$$E_x(z) = f\left(t \pm \frac{z}{v_p}\right)$$

where $n_p = \frac{1}{\sqrt{\mu_0 E}}$ and $f(t)$ is any twice differentiable function.

This solution is a wave propagating in the negative (plus) z direction if the plus (negative) sign is selected.

We consider some simple animations to provide some feeling for solutions in the optical domain. The first example is an pulse with a centre wavelength of 0.5 microns, in the green/blue region of the optical spectrum. The pulse oscillates at a centre frequency of at approximately 0.610^{15} Hz. This particular pulse has a Gaussian envelop. The pulse width of the Gaussian is 10 femtoseconds. The field amplitude of the pulse can be written

$$f(t) = \cos\left(0.610^{15}t\right)\exp\left(-\left(\frac{t}{10^{-14}}\right)^2\right).$$

The animation below shows the field amplitude as a function space along the axis of propagation. Each frame in the animation represents a time shift of 20 femtoseconds. The pulse propagates to the right at a velocity of 0.3 microns/ femtosecond.

The solution $f(t)$ need not be spatially confined. For Fourier analysis of fields, unbounded harmonic solutions such as $f(x)$ $= \cos\left(0.610^{15}t\right)$ are popular.

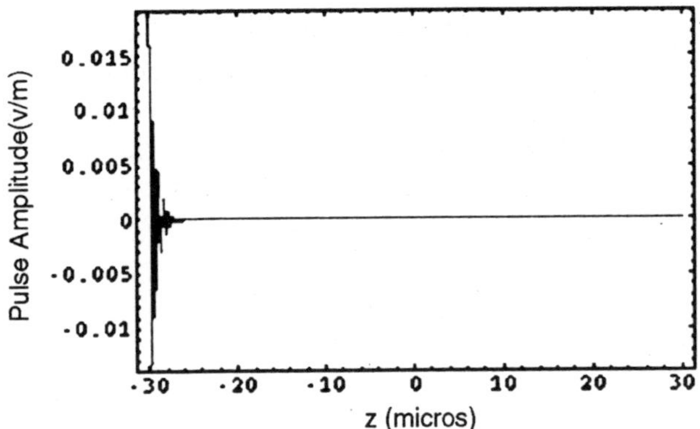

For such fields, one perceives wave propagation in the phase of the excitation. An animation of this field propagating in the plus z direction is shown below:

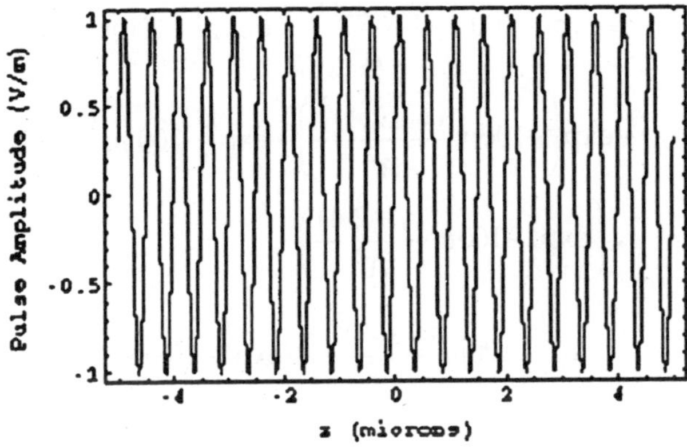

OPTICAL SENSORS

As mentioned in the introductory note on coherence theory, the nature of optical detectors plays a key role in the statistical analysis of optical systems. Detection of spatially distributed fields relied exclusively on photochemical materials or the photoelectric effect until the invention of CCD arrays in the mid'70s.

ANALYSIS OF FRESNEL DIFFRACTION

This note shows a couple of examples of the calculation of Fresnel transformations. Calculation of Fresnel diffraction presents a number of interesting challenges, some are pointed out by the example here, some you will discover on your own. Given that you are likely to encounter erroneous results, the important thing is to keep your intuition working.

Our first example is based on the input field

$$f(x,y) = \exp\left(-pi\frac{\left(x^2+y^2\right)}{A^2}\right)$$

Consider

$$f(x,\, y) = \left(\exp\left(-100pi\frac{(x+2A)^2}{A^2}\right)\right.$$

$$= \exp\left(-100\;pi\frac{(x+2A)^2}{A^2}\right) + \exp\left(-100pi\frac{(x)^2}{A^2}\right)$$

$$\left. + \exp\left(-100pi\frac{(x+2A)^2}{A^2}\right) + \exp\left(-100\;pi\frac{(x+2A)^2}{A^2}\right)\right)$$

$$\times\left(\exp\left(-100pi\frac{(y+2A)}{A^2}\right)\right.$$

$$+ \exp\left(-100pi\frac{(y+A)^2}{A^2}\right) + \exp\left(-100pi\frac{(y)^2}{A^2}\right).$$

$$\left. + \exp\left(-100pi\frac{(y+A)^2}{A^2}\right) + \exp\left(.-100pi\frac{(y-A)^2}{A^2}\right)\right)$$

NUMERICAL ANALYSIS OF COHERENT IMAGING

Coherent Imaging

The transfer function for a coherent imaging system is

$$\tilde{H}(\mu\nu) = P(-\lambda d_i\mu, -\lambda d_i\nu),$$

where $P(x,y)$ describes the transmittance of the output pupil of the imaging system. The impulse response of the imaging system is the impulse response of the transfer function:

$$\widetilde{h}(x_i y_i) = \iint (\lambda d_i \mu, \lambda d_i \nu) e^{j2\pi(x\mu + y\nu)} d\mu d\nu$$

We find the image plane field, $g(x_i, y_i)$, from the object plane field, $f(x_o, y_o)$, using

$$g(x_i y_i) = \int\int \tilde{h}(x_i - \tilde{x}_0, y_i - \tilde{y}_0) \frac{1}{M^2} f\left(\frac{\tilde{x}_0}{M}, \frac{\tilde{y}_0}{M}\right) d\tilde{x}_0 d\tilde{y}_0$$

where

$$M = \frac{d_i}{d_0}.$$

INCOHERENT IMAGING

Incoherent imaging is a linear system in the input intensity. The impulse response for an incoherent imaging system is equal to the squared modulus of the field impulse response for the corresponding coherent system. The transfer function for the incoherent imaging system is thus the autocorrelation of the coherent transfer function. Coherent imaging the transfer function for a coherent system described by the exit pupil function $P(x,y)$ is

$$\tilde{H}(u, \nu) = P(-\lambda d_i u, -\lambda d_i \nu)$$

The first section of this document analyses the incoherent transfer function and impulse response for various apertures.

Incoherent Transfer Functions and Impulse Responses

The normalized transfer function for an incoherent system is called the OTF, or optical transfer function. The Fourier transform of the OTF is the point spread function (PSF) of the imaging system. The OTF satisfies the following constraints:

1. $\mathcal{H}(0,0) = 1$
2. $\mathcal{H}(-f_x, -f_y) = \mathcal{H}^*(f_x, -f_y)$
3. $|\mathcal{H}(-f_x, -f_y)| \leq |\mathcal{H}(0,0)|$

Depth of Field

The impulse response for coherent image away from the focal plane is

$$h(x_i,y_i,x_0,y_0)=\frac{1}{\lambda^2 d_0 d_i}\exp\left[\frac{j\pi}{\lambda d_i}\left(x_i^2+y_i^2\right)\right]\exp\left[\frac{j\pi}{\lambda d_i}\left(x_0^2+y_0^2\right)\right]$$

$$\times\iint P(x,y)\exp\left[\frac{j\pi}{\lambda d_i}\left(\frac{1}{d_i}+\frac{1}{d_0}-\frac{1}{f}\right)\left(x^2+y^2\right)\right]$$

$$\times\exp\left[-\frac{j2\pi}{\lambda}\left(\frac{x_i}{d_i}+\frac{x_0}{d_0}\right)x\right]\exp\left[-\frac{j2\pi}{\lambda}\left(\frac{y_i}{d_i}+\frac{y_0}{d_0}\right)y\right]dxy$$

For the incoherent imaging system, the impulse response is simply $|h(x_i,y_i,x_0,y_0)|^2$. Squaring eliminates the phase factors in front of the integral and makes the impulse response shift invariant under the coordinate transformations on coherent imaging. The effective coherent transfer function for this transformation is

$$\tilde{H}(u,v)=P(-\lambda d_i u,-\gamma d_i v)\exp\left[j\pi\lambda_i^2\left(\frac{1}{d_i}+\frac{1}{d_0}-\frac{1}{f}\right)\left(u^2+v^2\right)\right]$$

Using this transfer function in the code above, we can put in a variety of values for d_i (assuming that d_o is fixed at 1 cm and f is fixed at 5 mm). The animation below shows the point spread function for a 1 mm aperture as the image position moves 100 microns toward the lens from the focal plane.

DIFFRACTION FREE BEAMS

Diffraction free beam is invariant with respect to diffraction from one plane to the next. A plane wave is an example of such a field. It turns out that plane waves are not unique in this respect and that unbounded solutions to the Maxwell equations exist which localize their energy along the optical axis more effectively than plane waves. Of course, it is not possible for a beam passing through a finite aperture to avoid diffraction.

The Bessel beam fields described here propagate without diffraction through unbounded apertures, but the energy of these beams is not finite and the beam does diffract after passing through an aperture.

One can easily verify that the field

$$E(\vec{r}) = e^{-j\beta} \int_0^{2\pi} A(\phi)\exp[j\alpha(x\cos\phi + y\sin\phi)]d\phi$$

where A(ϕ) is an arbitrary complex function of ϕ and $\beta^2 + \alpha^2 = \left(\frac{\omega}{c}\right)^2$, is a solution to the Helmholtz equation. The particular solution $A(\phi)=\frac{1}{2\pi}$ can be integrated to yield the non-diffracting beam

$$E(\vec{r}) = e^{-j\beta} \int_0^{2\pi} \exp[j\alpha(x\cos\phi + y\sin\phi)]\frac{d\phi}{2\pi}$$

$$= e^{-j\beta} J_0(\alpha\rho).$$

To see why this is a nondiffracting beam, consider its representation in k-space. The field consists of plane wave components drawn from a ring around the optical axis, as sketched below. Since each plane wave component experiences the same phase delay along the axis as the field propagates, all the components stay in phase and the transverse structure of the field is unchanged with propagation.

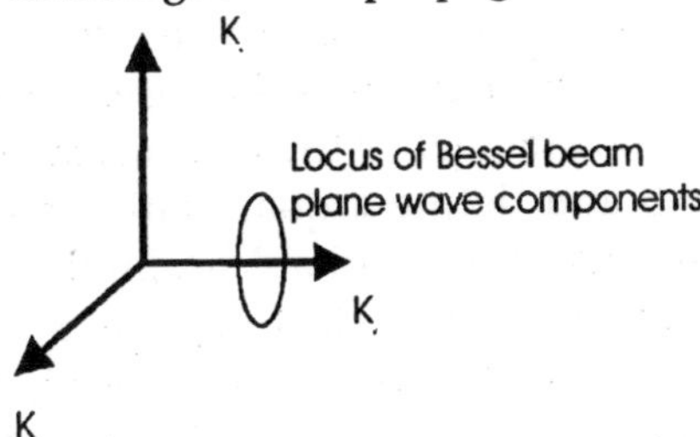

It is not necessary to use the entire ring to achieve this objective, you might want to consider fields drawn from partial segments on your own. Two points on opposite sides of the ring, for example, would form a nondiffracting cosine transverse pattern.

OPTICS AND INFORMATION

Of all physical phenomena, optics allows by far the highest capacity, greatest bandwidth and most effective means of information transfer. An optical fiber can transfer information at bandwidths 3-4 orders of magnitude beyond the fastest copper wire and optical imaging systems surpass the bandwidths and parallelism of ultrasonic and radio wave imaging systems by even greater margins.

Given these vast information transfer capabilities, analysis of information capacity and flow often lies at the heart of optical system analysis. This note briefly reviews the quantitative concept of information and briefly describes how it arises in optical analysis.

QUANTITATIVE MEASUREMENT OF INFORMATION

Information is stored in systems, transferred over channels and detected by measurements. A system might be "my crazy old aunt Beatrice, sitting on a sofa in the living room." Let's call this system S. Information from S might vary from details such as "Aunt Bea's knees are dirty" to "there is a nanogram of plutonium on the third gray hair back from Aunt Bea's left earlobe." Leaving aside the philosophical question as to whether or not system S stores an infinite amount of information, the quantity of information about S which can be transferred over a channel or detected by a measurement is always finite. Given that this is the case, we often consider the communication channel or the measurement device to be part of system S and define a quantitative measure for the information capacity of S. A suitable definition is:

The information capacity, H, of a system S is the minimum number of bits necessary to completely specify the state of S.

The number of yes/no questions that could be answered about S is an upper bound on H. For example the answer to the question "Are Aunt Bea's knees dirty?" could convey one bit of information.

The number of yes/no questions is an upper bound on H because the answers to some questions may be correlated or implicit in the nature of the system. For example, if Aunt Bea's knees are always dirty, the answer to the question may convey no information.

The definition of the information capacity makes comparisons between systems possible and makes information a universal concept. For example, if we can fully specify the state of S using one megabit, than the information capacity of S is equivalent to a one megabit computer memory and the systems themselves are in some sense equivalent. It is important to stress, however, that there is a difference between specifying the state of a system and describing the state of a system. The number of bits needed to fully describe Aunt Bea sitting on the sofa may greatly exceed the number needed to specify the state, particularly if describing Aunt Bea requires the use of poetry.

INFORMATION IN ANALOG SYSTEMS

In digital systems, the quantity information is equal to the number of bits available and is thus quite precise. In analog systems, such as an imaging system, information is more difficult to quantify. Noise and measurement resolution must be considered in analog systems and play key roles in determining information capacity. In analyzing the information capacity of imaging systems, it is often desirable to separate the impact of noise and measurement resolution, which are determined by the illumination source and the detector, from the impact of diffraction and spatial apertures.

Analysis of field propagation (diffraction) and boundary conditions (apertures) determines the number of electromagnetic modes which propagate through an imaging system. The temporal bandwidth and noise characteristics of the illumination source and the detector and digitization resolution on detection determine the quantity of information each mode can carry.

In analyzing the field in imaging systems, one focuses on the number of degrees of freedom. Each degree of freedom is a variable on which information can be encoded. For example, the magnitude and phase of a particular electromagnetic mode might constitute a degree of freedom and the number of available modes might constitute the number of degrees of freedom. Discussions of degrees of freedom allow one to estimate the potential information capacity of an analog system without fully considering the impact of noise and measurement.

A system with N degrees of freedom in which each degree of freedom could be independently placed and measured in r different states has an information capacity of

$$H = \log\left(r^N\right) = N\log(r)$$

Of course, it is critically important that the noise and measurement processes be such that r is greater than 2. In general, r is rarely large enough to play a major role in increasing the information capacity. Since the information is linear in N and linear in the log of r, it is almost always more useful to increase N rather than r.

As mentioned above, analysis of the degrees of freedom of an optical system may be considered from an electromagnetic perspective in terms of the modes of the system. A *mode* is a field distribution which satisfies both the Maxwell equations and the boundary conditions for a particular geometry.

Usually, one chooses to work with a set of modes which are *orthogonal,* which means that their inner product is zero, and which are complete, which means that any field satisfying the Maxwell equations in the geometry of interest can be described by a superposition of modes drawn from the set. In geometries where the field is confined to a limited range of space or spatial frequencies, a set of orthogonal and complete modes is generally discrete. In this case, the spatially continuous field distribution can be represented by a discrete (although potentially infinite) list of mode amplitudes.

THE SAMPLING THEOREM

Similar methods for representing continuous functions using discrete lists arise in linear systems analysis in the context of sampling theorems. Consider as an example a square integrable bandlimited function $f(x,y)$. We focus on 2D functions because our main focus will be on fields measured across 2D planes, such as image planes. By *bandlimited,* we mean that the Fourier transform of f is vanishes outside a certain range. We might express f as

$$f(x,y)=\int_{-B_{,}}^{B_1}\int_{-B_{,}}^{B_1}F(u,\mathrm{v})\exp[j2\pi(ux+\mathrm{v}y)]du d\mathrm{v}.$$

The Fourier transform of f, $F(u,v)$, is limited to a square region of area $4B_xB_y$ in the Fourier plane. A density plot of an example form of $F(u,v)$ is sketched below:

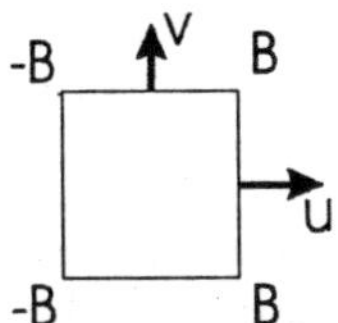

Consider a function $F_p(u,v)$, which is a periodic tiling of the Fourier plane with copies of $F(u,v)$, as sketched below.

Since $F_p(u,v)$ is a periodic function, it can be expressed as the Fourier series

$$F_p(u,\mathrm{v})=\sum_{\pi=-\infty}^{\infty}\sum_{\pi=-\infty}^{\infty}a_{\pi m}\exp\left[-j\pi\left(\frac{nu}{B_x}+\frac{m\mathrm{v}}{B_y}\right)\right].$$

Over the range of $F(u,v)$, $F(u,v) = F_p(u,v)$. This means that

$$f(x,y)=\int_{-B_x}^{B_1}\int_{-B_y}^{B_1}F_r(u,\mathrm{v})\exp[j2\pi(ux+\mathrm{v}y)]du d\mathrm{v}$$

$$=\sum_{x=-\infty}^{\infty}\sum_{m=-\infty}^{\infty}a_{nm}\int_{-B_x}^{B_x}\int_{-B_y}^{B_y}\exp\left[j2\pi\left(u\left(x-\frac{n}{2B_x}\right)+\mathrm{v}\left(y-\frac{m}{2B_y}\right)\right)\right]du d\mathrm{v}$$

$$=\sum_{x=-\infty}^{\infty}\sum_{m=-\infty}^{\infty}a_{nm}\operatorname{sin}c\,(2B_xx-n)\operatorname{sin}c\,(2B_yy-m)$$

where

$$\sin c(x) = \frac{\sin(\pi x)}{\pi x}$$

Considering the case $x = \frac{n}{2B}, y = \frac{m}{2B_y}$, we find $f\left(\frac{n}{2B_x}, \frac{m}{2B_y}\right) = a_{nm},$

which yields the Whittaker-Shannon sampling theorem,

$$f(x,y) = \sum_{x=-\infty}^{\infty} \sum_{x=-\infty}^{\infty} f\left(\frac{n}{2B_x}, \frac{m}{2B_y}\right) \sin c(2B_x x - n) \sin c(2B_y y - m)$$

The sampling theorem means that the continuous function $f(x,y)$ can be fully represented by the discrete set of samples

$$f\left(\frac{n}{2B_x}, \frac{m}{2B_y}\right)$$

In cases of particular interest, we focus on the value of $f(x,y)$only over some finite aperture. If we leave $f(x,y)$ unconstrained outside this aperture, only a finite number of samples is needed to characterize the function. If the extent of the aperture of interest is (-X,X) and (-Y,Y), the number of samples needed is $16B_xB_yXY$. This number of samples corresponds to the number of degrees of freedom in the system "$f(x,y)$ over the aperture of interest." The information capacity of this function is proportional to this number of degrees of freedom.

Geometrical optics: The domain in which optical waves can be treated as simple straight-line rays and the effect of optical elements can be modeled using simple geometrical constructions. There are a number of requirements for these approximations to be an accurate description of an optical system.

The most important is that $\pi/D \ll 1$, where π is the wavelength of the light signal and D is the size of the object, lens, aperture, etc. that is being studied. This requirement is almost always satisfied for visible light interacting with everyday-sized objects (which is why Newton did not find a contradiction from everyday experience when he stated that light was probably a stream of particles).

There are a number of other requirements that must be fulfilled in some cases. These requirements will become important in the studies of lenses and curved mirrors.

PROPERTIES OF SHADOWS

Shadow: produced on a screen behind the object when light from a source is blocked (either completely or only partially) by the object. In the case where geometrical optics is appropriate, the shadow can be constructed using simple straight-line rays.

A clear unambiguous shadow requires a "point" source of light – something that either is very small compared to the other dimensions of the experiment or else is very far away. The geometrical construction of the shadow is particularly simple in this case, since all of the rays that strike the object come from a single point, so that a given point of the image either receives light or doesn't.

If the source of light is too large to be regarded as a "point" source, then the shadow is constructed using the principle of superposition – the extended source is modeled as a large number of point sources, and the shadow produced by each of these point sources is constructed independently. This is an example of the "principle of superposition" – that is that the effect of a number of elementary sources is the sum of the effects of each one of them taken individually.

The shadow produced by two or more point sources is also handled by the principle of superposition in the same way – the light striking a screen behind the object is the sum of the contributions from all of the light sources taken individually.

The *umbra*: the portion of the shadow which is totally dark because light from all sources is absent.

The *penumbra*: the portion of the shadow which is illuminated by some of the sources and is therefore not completely dark.

In some situations (such as a transparency projector or an x-ray picture), the information content of the image is really in the shadows, and the unobstructed parts of the image are not of interest. False shadows are often used to make a flat image appear "3-dimensional." This works because our eyes are accustomed to the idea that a shadow is produced when an object is in front of a screen.

SOLAR AND LUNAR ECLIPSES

Since the Moon orbits the earth with a period of about 29.5 days, there are occasions when the Moon comes between the Earth and the Sun or when the Earth comes between the Moon and the Sun. These are called eclipses, and they can be understood using the principles of shadows outlined above. Although the Moon is much smaller than the Sun, it is much closer to the Earth, so that its shadow can totally block the light of the Sun if the alignment is exactly correct. This is a "total" eclipse; it is quite rare, since the alignment is usually not exact. The more usual situation is a "partial" eclipse in which only part of the Sun is blocked. Even during a total eclipse, the umbra of the Moon does not completely cover the Earth, so that only some places on the Earth see the eclipse as total.

Other locations see only a partial eclipse, and yet other locations see nothing at all, since they are neither in the umbra nor in the penumbra. A lunar eclipse occurs when the Earth comes between the Sun and the Moon. As above, the eclipse may be total or partial, depending on the exact alignment. Since the Moon shines by reflecting light from the Sun, the moon almost disappears during an eclipse.

However, even in a total eclipse, it is often possible to see a very faint image of the Moon because some light from the Sun is deflected by the atmosphere and still reaches the Moon.

THE PINHOLE CAMERA

THE RELATIONSHIP BETWEEN AN OBJECT AND ITS IMAGE

An object can be anything that we observe using an optical system. If it is a physical object, then we imagine that light rays are emitted in all directions from every point on the object. The intensity (and the colour) of these rays may vary from point to point. An object may also be virtual. A virtual object is usually the image of a physical object formed by an optical system. An exámple would be the person that looks back at us when we look in a mirror.

If an optical system forms an image, that means that there is a one-to-one relationship between a point of the object and a point of the image. That is, a point of the image receives light from only one point of the object. The object and the image need not be the same size or have the same orientation. The requirement is only that any given point of the image receive light from only one point of the object.

This one-to-one relationship is usually not realised when light from an object is simply incident on a screen or other target. A single point on a screen usually receives light from many different points on an object, so that there is no unique one-to-one relationship between object and the light that strikes a point on the screen.

We see distinct objects because the lens at the front of our eyes produces an image of the scene on the retina – the light-sensitive "screen" at the back of the eye. This image is really our only connection to the real world, and we can be fooled in interpreting this image if the assumptions that we unconsciously make in evaluating it are not true. This is the basis for "optical illusions," for our ability to see three-dimensional effects in two-dimensional pictures, etc.

THE PINHOLE CAMERA

The pinhole produces the one-to-one relationship between the object and the image because of its small size. The image is dim because most rays from the object can't contribute to the image The image is inverted (both with respect to up/down and right/left) – a feature of many imaging systems

Advantages: cheap and simple; can be used for any wavelength; objects at any distance produce clear images (great "depth of field") Pinhole approximation gets better as hole gets smaller Image also gets dimmer – fewer rays can get through Watch out for λ/D if pinhole gets too small – requirement for geometrical optics may not be true if pinhole is small enough

There is a simple relationship between the size of the object and the size of the image. This relationship can be seen from the geometrical picture shown below. The object is an arrow on the left that points upward. The image is constructed by drawing straight lines from points on the object to points on the imaging screen, which is shown to the right of the pinhole. Because the pinhole is so small, only one ray from each point on the object reaches any point in the image plane.

The two triangles formed by the object and its two rays and the image and its two rays are similar, and the sides are therefore proportional. If the lengths of the object and image are L_0 and L_i respectively, and if the corresponding distances between the pinhole and the object and image are D_0 and D_i respectively, then using the properties of similar triangles: Magnification= $L_i/L_0 = D_i/D_0$

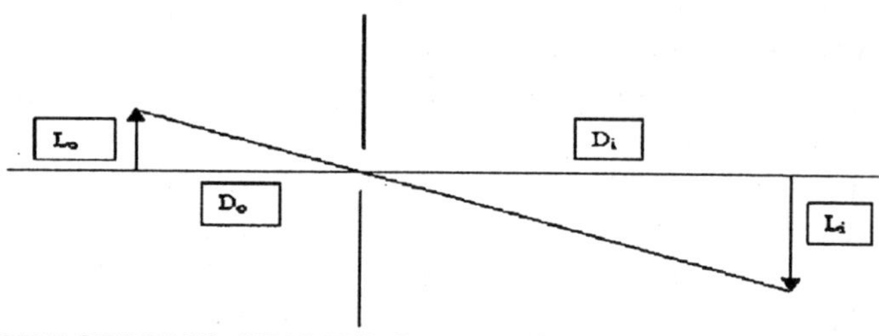

REFLECTIONS, WAVES ON A ROPE

Model a rope as a series of small masses connected by rubber bands. When one of the masses is displaced from its equilibrium position, it exerts forces on its neighbours. These forces cause the adjacent masses to move and also act to bring the initial mass back to its equilibrium position. If all of the masses and rubber bands are the same then the displacement propagates from one mass to the next one. In the absence of friction or other loss mechanisms, the resulting wave propagates without any decrease in its amplitude – each mass transmits its energy to the next mass in the chain through the rubber-band link between them.

The propagation of the wave consists of an exchange between two types of energy: the elastic potential energy that is stored in the stretched rubber bands is converted to kinetic energy of the moving masses, and this kinetic energy in turn is converted back into elastic potential energyy when the next rubber band is stretched.

This is analogous to a light wave which propagates by an exchange between electric and magnetic potential energies. The velocity of propagation in this case depends on two quantities:

- The stiffness of the rubber bands, which determines the magnitude of the force that is transmitted to an adjacent mass for a given displacement,
- The size of each mass, which determines how quickly the mass will respond to the transmitted force. The velocity of propagation varies directly with the stiffness of the rubber bands and inversely with the mass at each junction.

When the disturbance reaches a fixed object (or a mass that is much greater than any of the others), the same force produces a much smaller displacement (possibly no displacement at all if the object is fixed), so that the restoring force is greater than at any other point on the rope. The result of the larger restoring force is to generate a reflected wave with a displacement that has the opposite sign. Using the previous paragraph, a larger mass implies a smaller propagation velocity.

When the disturbance reaches the free end of the rope (or a mass that is much smaller than any of the others), the same force produces a much larger displacement than at other points so that a reflected wave of the same sign is generated.

In the more general case, a reflection is generated whenever a wave reaches a point where its velocity of propagation changes. The reflected wave is reversed when the velocity decreases at the boundary and is not reversed when the velocity increases. Using the various definitions of the phase of a wave, the reversal at the reflection can be thought of as a phase change of one-half of a wavelength or as 180°ð in a sine-wave cycle or as a time delay of one-half of the period of the wave.

Our eyes and many other optical detectors are sensitive only to the amplitude of the light wave (or to the intensity, which is proportional to the square of the amplitude) and not to its phase, so that these reversals produce no visible change in the light and can be ignored. However, this is not always the case, and the phase-reversal on reflection plays an important role in interference between two light waves.

REFLECTIONS, RADAR AND SONAR

If the velocity of propagation of a wave is known, the time it takes the reflected wave to return to the source is a measure of the distance between the source and the object that produced the reflection. This is the basis for radar, sonar and time domain reflectometry (TDR), all of which measure the distance between a transmitter and a reflector using the time it takes the echo of a pulse to return to the sender.

In order for radar and sonar to be useful, the system must operate in the limit where geometrical optics is valid. In other words, these methods will only "see" an object if its size is considerably larger than the wavelength that is used, so that the ratio of λ/D is less than 1. For example, if a bat is to detect an insect whose size is about 6 millimeters in diameter (about 0.25 inches), the bat must use a sound wave whose wavelength is significantly smaller than this value. The velocity of sound in air is about 335 m/s. If we take a wavelength of 3 mm as just barely satisfying the criterion of being not larger than the object that is to be detected, the frequency of the sound wave must be at least $\nu = c/\lambda = 335/3 \times 10^{-3} \sim 112$ kHz

This is in fact about the frequency that bats use. It is much higher than we can hear. Shorter wavelengths (that is, higher frequencies) allow the bat to detect smaller objects or to see the structure of a larger object.

Note that we use the symbol "c" for the velocity of sound in air (rather than "v") to avoid confusing it with the frequency ν.

Time domain reflectometry is often used to detect the location of a broken power cable in a remote region. When a pulse of voltage is applied to one end of the cable, the broken end of the cable causes a reflection to be sent back to the transmitter. The time it takes the pulse to return gives the approximate location of the break. The same technique is used to detect the location of a break in an underground telephone cable.

Unlike our eyes or our cameras, which are simply receivers that detect objects using reflected light that is produced by some external source, radar and sonar systems also must include an active transmitter. They are therefore both more complex and more versatile – they can detect objects even in the absence of ambient light, but they must include a separate transmitter to make this possible.

The range of radar and sonar systems is limited by two factors. The maximum range is determined by the amount of power that the transmitter can produce and by the sensitivity of the receiver that is used to detect the echo. The minimum range is given by the timing resolution of the receiver – that is, by the ability of the receiver to distinguish between the original transmitted pulse and the (usually much weaker) reflected echo. When the range to the target is too small, the echo can arrive so quickly that the transmitted pulse has not yet died away.

The goals to be able to detect objects that are very far *or* very close to the transmitter using a single system conflict with each other to some extent. The maximum range of an echo system is increased by increasing the transmitted power, but this increase will saturate the receiver for a longer period of time and make it more difficult to detect an object that is very nearby. Radar systems have clever pulse shapes (called chirps) that attempt to minimize this conflict. Bats use the same techniques.

REFLECTIONS,METALS AND DIELECTRICS

The simple model of a metal is a regular, rigid arrangement of positively charged ions and a cloud of electrons that are relatively free to move inside the material. The free electrons respond to applied electric fields. Since the electrons are approximately free particles, they arrange themselves on the surface of the metal so as to cancel the electric field in the interior. For example, an external positive charge attracts electrons to the surface of the metal. This attraction continues to draw more electrons to the surface until the repulsion between these electrons at the surface exactly cancels the attraction of the external positive charge.

Since the electrons in a metal are free particles and the electric field does not penetrate into the interior of the material, very little of the incident energy is absorbed, and most metals are good reflectors. Ionized gases, which are also made up of positive ions and relatively free electrons behave very much like metals in this regard, and tend to be good reflectors as well. Atoms near the top of the atmosphere are ionized by ultraviolet radiation from the sun, and this layer of ionization is a good reflector of radio waves.

Most liquids do not contain free electrons, but they may contain positive and negatively charged ions, which can respond to an electric field. However, the ions are usually not as free as electrons are in a metal, and most liquids are more like dielectrics even if they have lots of charged ions.

The approximation of completely free electrons and passive, fixed ions that is used to model a metal breaks down for two reasons as the incident frequency is raised. The electrons are not really free particles and cannot respond infinitely rapidly to changes in the electric field. This means that they become unable to follow the field as its rate of change increases. In addition, the positively charged ions will interact with the incident field as well, especially when specific frequencies (which are characteristic of the particular ion are incident).

Thus metals tend to be reflectors at low frequencies, but absorbers as the frequency is raised. Some metals (gold, for example), absorb the blue light at the high-frequency end of the visible spectrum and appear yellow as a result. This effect is even more pronounced in copper, which absorbs most of the visible spectrum. Since it reflects only the lowest-frequency portion of the visible spectrum, it appears reddish-orange as a result.

The same thing is true for ionized gases – they tend to change from reflectors to absorbers as the frequency of the incident radiation is raised. If the density of the gas is not too great, there may not be enough atoms to absorb all of the incident radiation, and the gas may become approximately transparent as a result. Thus the ionosphere reflects radio waves but transmits signals in the visible portion of the spectrum. It becomes strongly absorbing for ultraviolet frequencies because these frequencies are strongly absorbed by oxygen and other atmospheric gases.

The simple model of a dielectric is a substance which may have the same regular structure as a metal, but with very few free electrons. The details of the absorption and reflection depend on the details of the material, and it is usually not possible to give a general rule as it was with metals. However, dielectrics generally transmit and reflect the incident energy with coefficients that are specific to the material and are usually functions of the incident frequency. This dependence on frequency is usually much more complicated than the relatively simple variation that characterizes metals and ionized gases.

MIRRORS

Since metals make the best general-purpose reflectors, they are usually used to make mirrors. Silver has a very high reflectivity across the entire visible spectrum and is usually the first choice for a mirror; aluminum is almost as good. Both of these metals interact with the oxygen and with the other components in the atmosphere to form compounds (mainly oxides and sulfides) that are not metallic and do not reflect nearly as well as the pure metal itself.

The metallic reflecting surface is therefore protected in some way. Silver mirrors usually have a pane of glass in front of them; aluminum mirrors are often coated with a thin layer of aluminum oxide. In both cases, the protective layer degrades the reflectivity somewhat but protects the underlying metallic surface from oxidation.

The simple model of a metal as a perfect reflector breaks down if the metallic layer is very thin (that is, only a few atoms thick), so that there is not much difference between the surface of the material, where the external electric field is concentrated and the interior, which we modeled as a field-free region. These mirrors are often called "half-silvered" because the thickness of the layer is adjusted so that about 50per cent of the incident light is reflected. Since very little light is absorbed, the other 50per cent must be transmitted, and these mirrors are often used as "one-way" mirrors.

If a one-way mirror is placed between two rooms that have very different levels of illumination, a person standing in the bright room cannot see through the mirror because the 50per cent of the light that is reflected on the bright side is greater than the 50per cent of the much lower level of illumination on the dim side. A person on the dim side, however, can easily see into the bright room because 50per cent of the light from the bright side is transmitted.

The same effect can be produced with ordinary window glass, even though it is not a 50/50 reflector. Window glass typically reflects about 4per cent of the light that strikes it, but even this small reflection may be significantly larger than the 96per cent of the light transmitted from the other side. Thus ordinary windows act as mirrors for people inside a house at night, while outsiders can easily see in. The reverse is true during the day, when the outside is much brighter than the inside. People inside the house can easily see out, whereas those outside cannot see in because they cannot detect the light transmitted from the interior in the presence of the 4per cent of the light outside the house that is reflected from the windows. The windows therefore look like mirrors to those outside.

WAVEFRONTS AND HUYGENS' PRINCIPLE

A wavefront is a surface over which an optical wave has a constant phase. For example, a wavefront could be the surface over which the wave has a maximum (the crest of a water wave, for example) or a minimum (the trough of the same wave) value. The shape of a wavefront is usually determined by the geometry of the source. A point source has wavefronts that are spheres whose centres are at the point source. A fluorescent tube would have wavefronts that are cylinders concentric with the tube itself. A very large sheet of material that is uniformly illuminated would generate wavefronts that are plane waves parallel to the sheet.

The direction of propagation of the wave is always perpendicular to the surface of the wavefront at each point. Thus, the wavefronts of a point source are spheres and the wave propagates radially outward – the radius of a sphere is perpendicular to its circumference at each point. The same thing is true of the radius of the cylindrical wavefronts that would be generated by a fluorescent tube.

Although the wave fronts produced by a point source are always concentric spheres in principle, when the source is very far away the radii of the spheres are so large that they look like plane waves to an observer. (Just as the Earth looks flat when viewed from a point near its surface.)

Huygens' principle describes how a wavefront moves in space. According to this principle, we imagine that each point on the wavefront acts as a point source that emits spherical wavelets. These wavelets travel with the velocity of light in the medium. At any later time, the total wavefront is the envelope that encloses all of these wavelets. That is, the tangent line that joins the front surface of each one of them. A simple example of how a plane wavefront moves is shown below:

The same construction is used for a wavefront of any other shape.When a wave travels in a single medium at a constant speed, the Huygen's construction preserves the general form of the wavefront. That is, spheres propagate and become larger spheres, cylinders become larger cylinders, etc.

If a portion of the wavefront enters a different medium (enters glass from air, for example), then the wavelets generated by each portion of the wavefront travel with the velocity that is appropriate for the medium that the wavefront is in. That is, the wavelets in the medium where the speed of light is less will have smaller radii than the wavelets in the original medium.

Although Huygens' principle was initially stated without any proof, a slightly modified form of it was later derived by Fresnel from the mathematical theory of waves. Note that Huygens was a contemporary of Newton, and that it would probably have been much more difficult to publish his theory if he had lived in England, where disagreeing with Newton was not an easy or popular position to take.

Mirrors, The Law of Reflection

When a light wave is incident on the boundary between two media, the "angle of incidence" is the angle between the incident rays and the normal to the surface at that point. The reflected ray emerges at the same angle with respect to the normal (the perpendicular to the surface at that point). The incident ray, the reflected ray and the normal all lie in the same plane. This result can be derived from Huygens' principle using a simple geometrical construction of the incident wavefront.

Since the wave is incident on the boundary between the two media at some angle other than 0 degrees or 90 degrees, one part of the wavefront reaches the boundary before the rest of the wavefront. The reflected wavelets from that part start radiating backwards while the remainder of the wavefront is still moving towards the boundary. The reflected wavefront can be completed when the last part of the incident wave reaches the boundary. A simple geometrical construction shows that applying Huygens' principle to this case results in the angle of incidence being equal to the angle of reflection.

If the reflecting surface is plane and smooth then a parallel beam of light produces a reflection whose rays are also parallel. This is "specular" reflection. If the surface is curved or not smooth, then the direction of the reflected beams will vary from point to point, and a parallel beam of light will produce a more complex reflection. In the limit of a reflecting surface that is very rough and irregular, the reflected light goes in all directions with more or less the same intensity. This is called "diffuse" reflection, in which the initial directivity of a parallel beam is completely lost. Note that this diffusion is due to the fact that the angle of incidence varies rapidly from point to point (even though the incident beam is composed of parallel rays) because the surface roughness means that the normal to the surface varies from point to point.

The law of reflection relating the angle of incidence to the angle of reflection is true for all materials, but the *amplitude* of the light reflected in a particular direction depends on the angle of incidence, on the type of material (metal or dielectric) and on the polarization of the incident radiation. This variation turns out to be much more important for reflection from dielectrics than for reflection from metals.

For most dielectrics, the amplitude of the reflected light increases as the angle of incidence increases, and most dielectrics become almost 100per cent reflectors at large angles of incidence (approaching 90°ð). Furthermore, the amplitude of the reflected light depends on the polarization of the incident beam, and there are certain angles where the reflected amplitude goes to 0 for one polarization so that only other polarizations are reflected.

When the road is dry, its surface is very rough, and the reflection is diffuse. Incident light from any direction is reflected in all other directions with essentially equal amplitude, so that light reflected from the road reaches our eyes and we can see it at night. When the road is wet, the surface becomes much more regular, and the reflection tends to be specular rather than diffuse. Light from an overhead street light tends to be reflected back upward and not into our eyes (the angle of incidence is quite small and therefore the angle of reflection is small as well).

Light from the headlights of the car tends to be reflected forward – both the angle of incidence and the angle of reflection are large in this case. Very little light is reflected back to our eyes both because reflection is specular. Since the reflection coefficient tends to approach 100per cent at large angles of incidence, oncoming traffic sees a large amount of light that is reflected from the roadway at large angles of incidence. Most of the light from the headlights is of almost no use in seeing the road and simply causes glare for the oncoming traffic.

Refraction and Snell's Law

When light strikes the boundary between two different media, a portion is reflected according to the laws of reflection, and another portion enters the medium. The ray which crosses the boundary and enters the second medium has been *refracted,* and the process at the boundary is called *refraction.*

The velocity of light depends on the properties of the medium it is traveling in, and especially on the density of the material. The velocity of light decreases when light enters a denser medium. The ratio of the speed of light in vacuum to the speed of light in any material is called the index of refraction of that material. If c is the speed of light in vacuum and v is the speed in some material, the index of refraction is $n=c/v$

Since v is always less than c, the index of refraction is always greater than 1. Note that the index of refraction is a ratio of two speeds and therefore has no units. Another way of expressing the same relationship is:$v=c/n$ That is, the speed of light in any medium is the speed in vacuum divided by the index of refraction.

The index of refraction depends on the detailed properties of the material and may also vary with external parameters such as ambient temperature, etc. However, we will generally ignore these variations and assume that the index of refraction is a constant that is a characteristic of the material. In addition, although the index of refraction of the atmosphere is about 1.0003, we will often take this index to be exactly 1, so that the speed of light in air is equal to its vacuum value. Using Huygens' principle, it is easy to show that when the velocity of light decreases at the boundary then the refracted light is bent towards the normal. That is, the angle between the refracted ray and the normal is smaller than the angle between the incident ray and the normal. Conversely, when light crosses a boundary into a medium where its velocity is greater, the angle of the refracted ray with respect to the normal is larger than the angle of incidence. This relationship is called *Snell's Law*; the mathematical form of the law relates the incident angle, the refracted angle and the indices of refraction of the two media at the boundary. The relationship actually involves the sines of the angles rather than the angles themselves. Since light that leaves a denser medium and enters a less dense one (glass to air, for example is bent *away* from the normal, the angle of refraction will always be greater than the angle of incidence in this case.

Therefore, the angle of refraction can exceed 90 degrees if the angle of incidence is large enough. There is no refracted wave at this angle, which is called the *critical angle* – the light is only reflected. This is called *total internal reflection*. There is also no refracted wave for all angles of incidence larger than the critical angle. This effect is the basis for optical fibers, which effectively "trap" light inside the fiber because of total internal reflection.

There are a number of optical effects that are caused by the variation of the index of refraction. A common effect is a mirage, which is caused by the variation of the index of refraction of air with temperature. The air near the surface of a roadway is often much hotter than the air at higher elevations, and its density and therefore its index of refraction are smaller. Light incident on this air from above is therefore bent away from the normal.

The light that finally strikes the road is almost perfectly reflected, since the angle of incidence is close to 90 degrees, and these reflected rays will appear to come from a virtual object that is below the surface of the roadway. Since this effect is common over water, our eyes often assume that the hot roadway is actually a body of water.

Dispersion, Prisms and Rainbows

The index of refraction of all materials varies with the frequency or wavelength of the incident light. This effect is called *dispersion*. Using quantum mechanics and electromagnetic theory, it is possible to show that this variation is a direct consequence of the fact that atoms and molecules absorb and emit only discrete frequencies or wavelengths.

·In general, the index of refraction for most materials in the visible portion of the spectrum increases with increasing frequency. Thus the index of refraction for blue light is larger than for red light. This is not a fundamental principle of physics – it just happens to turn out that way. In fact, there are a number of materials where the inverse is true – it just happens that these situations are not usually observed in everyday experience.

Since the speed of light is equal to the vacuum speed divided by the index of refraction, the increase in the index from red to blue means that blue light travels more slowly than red light in most materials. In addition, the variation of the index with frequency or wavelength means that blue light is bent more than red light when a light beam crosses the boundary between two media.

This was first observed by Newton.Angular dispersion, in which different wavelengths are bent through different angles at a boundary, and temporal dispersion, in which different wavelengths travel at different speeds in most media, play very important roles in everyday experience and in technology. Angular dispersion is what causes a rainbow, for example, and temporal dispersion is widely used in satellite communications systems to measure the properties of the ionosphere and the atmosphere.

Real and Virtual images; Plane Mirrors

A real image is produced on a screen (or some other detector) when all of the rays from a single point on an object strike a single point on the screen. The intensity distribution on the screen is a reproduction of the intensity distribution emitted by the object. The image may be smaller or larger; it may be erect or inverted – it is the one-to-one relationship between the image and the object that is important.

A virtual image is produced when rays of light reach our eyes that appear to come from a real object, but there is in fact no object at the apparent source of the light. The most common example is when light from an object strikes a simple plane mirror. The reflected rays appear to come from an identical object that is located behind the mirror. The spatial distribution of rays is completely consistent with what we would see if there really was an object behind the mirror. Note that we cannot actually place a screen at the point where the image appears to be.

The law of reflection – that the angle of incidence is equal to the angle of reflection, can be used to construct the image from any object placed in front of a mirror. This construction is relatively simple when the mirror is a single flat surface. A ray from object point O that strikes the mirror results in a reflected ray that appears to be coming from image point I.

This construction is the same for every ray from O that hits the mirror anywhere, so that there is a one-to-way relationship between the points O and I. However, point I is behind the mirror, and is therefore a virtual image. However, the rays reflected by the mirror are completely consistent with those that would be produced if there were a physical object at point I.

This construction can be completed for any point on the object side of the mirror, so that the image space is an exact reproduction of the points on the object side of the mirror. The reflected rays preserve the size and shape of the object, so that the magnification (the ratio of the size of the image to the size of the object) is point I always appears to be the same distance behind the mirror as the object point O is in front of it.

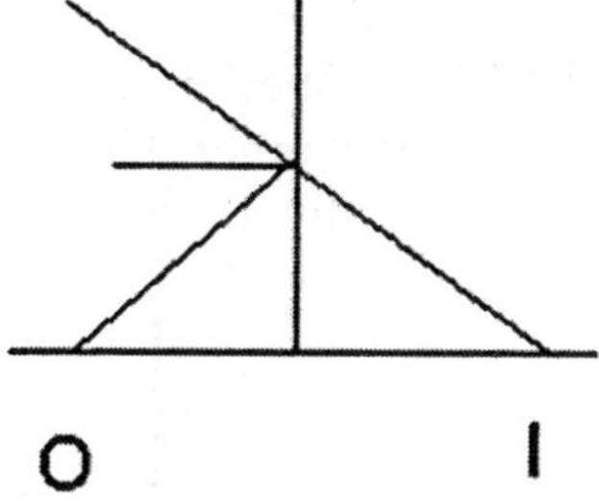

The mirror has therefore reversed the coordinate axis perpendicular to the plane of the mirror – object points that are further to the left of the mirror in the image points that appear to be further to the right in the virtual image.

If the geometrical construction above is repeated for a point just above point O, the resulting image point will be just above point I. Thus the mirror preserves the vertical relationship between objects.

An object O even if the mirror does not extend that far, since the reflected ray only uses a higher point on the mirror. For example, if the object O is your shoe, then you can see the reflection of your shoes with a mirror that is only one-half of your height.Finally, these constructions are exact – the image is a perfect replica of the object (within the limits imposed by geometrical optics). This is the last time in our study of mirrors and lenses that we will be able to say that.

In addition to the single-mirror geometry shown above, there are a number of useful configurations of two (or more) mirrors. Two of these configurations are shown in the following diagrams:

- Two parallel mirrors (or any two parallel surfaces) produce an output beam that is offset in space but parallel to the input:

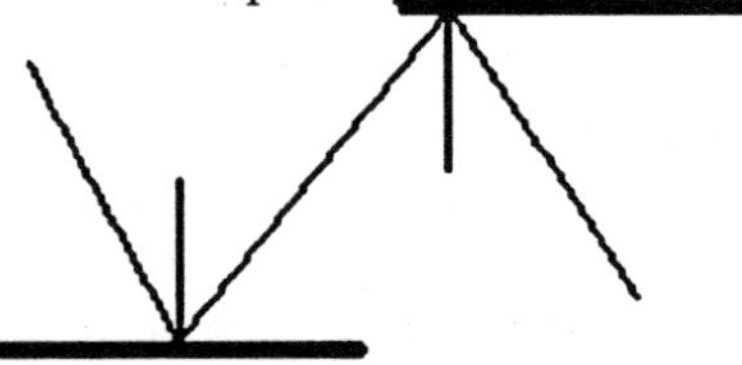

- Two mirrors at 90 degrees (a "corner cube") reflect a light beam back along its direction of incidence, no matter what the direction was.

These constructions are special cases of two mirrors that are tilted with respect to each other. The output beam is offset by 360 degrees minus twice the tilt angle.

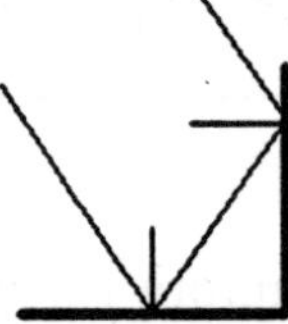

In general, if a mirror is tipped by some angle, then both the angle of incidence and the angle of reflection change by this tip angle. The angle between the incident and reflected beams therefore changes by twice the tilt angle.

FIELD OF VIEW

The field of view is the size of the largest object that can be seen through an optical system by a detector that is located at a fixed position. The size of the field can be expressed in units of length (such as meters), although it is more convenient to use an angular measure (e.g., degrees) in some situations.

For example, if you are looking at your own image in a plane mirror, then both the object (your shoes, for example) and the detector (your eye) are the same distance from the mirror.

Although the principles are unchanged, the detailed situation is somewhat more complicated when the object and the detector are not at the same distance from the mirror. For example, consider the field of view of the side mirror in a car.

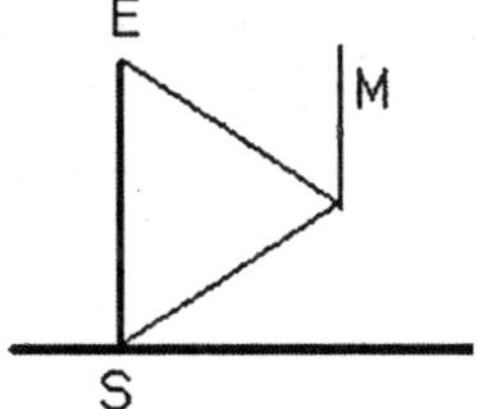

Since the two extreme rays are not parallel to each other, the field of view is a pie-shaped region bounded by the two rays that are shown. This region can be described either in terms of the wedge angle of the pie or in terms of its width at some distance behind the car.

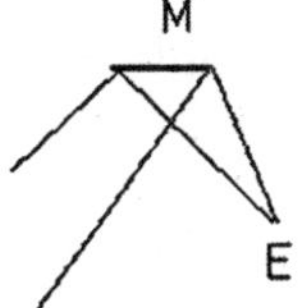

The mirror in a typical car is about 16 cm wide (about 6.5 inches), and the distance from your eye to the mirror is about 90 cm (35 inches). The field of view of this configuration is a pie-shaped region where the wedge angle of the pie is about 10 degrees. If we extend this pie-shaped region to just behind the back of the car (about 25 feet or 7.5 meters behind the driver), the width of the pie at that point is about 1.3 meters (about 4.5 feet).

This pie-shaped region can be moved by changing the position of the mirror, which changes the angles of incidence of both rays simultaneously. However, the wedge angle of the pie and the width of the region that can be viewed remains essentially fixed, because these parameters are set by the width of the mirror and the distance between the mirror and the driver's eye, and these parameters do not change when the mirror is tilted.

The region that can be viewed is less than the width of a typical lane on a roadway, so that this mirror will always have a blind spot– a region where another vehicle might be located but which cannot be seen by the driver. Rays emitted by objects in that region strike the mirror, but they cannot reach the driver's eye because there is no point on the mirror where the angle of incidence of the ray produces a reflected ray which is directed towards the driver's eye.

Spherical Mirrors

The principle that the angle of incidence is equal to the angle of reflection is also applicable to a mirror that is curved. The most common example is a spherical mirror, although other shapes (such as a mirror in the shape of a parabola) are also used. Since a line along the radius from the centre of the sphere to the circumference is always perpendicular to the circumference, every radius line can be used to construct the normal to any point on the sphere.

When an incoming ray (such as "I") is incident at some point on the sphere, it is reflected as ray R using the usual rule that the angle of incidence is equal to the angle of reflection.

Although the direction of the radius vector changes from point to point, all rays that are parallel to ray I are reflected in such a way that the reflected rays appear to come from a single point. This is called the focal point and is midway between the centre of the sphere and the circumference. Unlike a plane mirror, where all of the rays really did appear to come from a single point, the situation here is only an approximation that is reasonably correct provided the rays do not strike the sphere too far away from the axis.

Since the angle of incidence and the angle of reflection are always equal, the law of reflection is symmetric, and a ray that was incident along direction R would be reflected along direction I. Thus we can formulate the 3 rules that describe the reflection of light rays from a spherical mirror:

- A light ray that is incident along a direction parallel to the axis of the mirror is reflected so that it appears to come from the *focal point* – a point that is halfway between the centre of the mirror and the circumference along the axis.
- A light ray that is incident along a direction that is pointed directly at the focal point is reflected parallel to the axis. This is the same as the previous rule with the roles of the incident and reflected rays interchanged.
- A ray that is incident along the radius vector is reflected back on itself. Such a ray has an angle of incidence of 0 by definition, and its angle of reflection is therefore 0 as well. This in the only one of the 3 rules that is exactly true.

In addition to the three common rules, there is a 4th rule that is often useful. Since the mirror is always symmetric about its central axis, a ray that strikes the mirror at its centre (at any angle of incidence) is reflected symmetrically backwards by the same angle below the axis.

It is important to remember that rules 1 and 2 are approximations that are valid only when the incident rays are not too far from the axis. This is called the *paraxial ray* approximation. This approximation is an important limitation for reflections and refractions at curved surfaces that are spherical. (It is possible to remove this limitation in some cases using non-spherical shapes, but these shapes are usually too expensive to fabricate and are used only in very special situations.)

The validity of the *paraxial ray* approximation is determined by how far the incident ray is from the axis. The scale factor is the ratio of the distance of the ray from the axis divided by the focal length, and the paraxial ray approximation requires that this ratio be small compared to 1. This ratio often appears as its reciprocal: the ratio of the focal length to the size of the object (or to the diameter of the lens or the mirror), and the paraxial ray approximation implies that this reciprocal ratio be much larger than 1. This reciprocal ratio of the focal length to the size of the object or the diameter of the lens or the mirror is called the f-number, and it plays an important role in many other aspects of the formation of images by lenses and mirrors.

Using this notation, the paraxial ray approximation is equivalent to the statement that this simple model of constructing the image produced by a lens or mirror is valid only when the f-number is large, and the approximation becomes increasingly poor as the f-number decreases.

The following diagram shows the rules governing the reflection of rays from a convex, that is one whose curvature is outward towards the source of the rays. Note that the fact that the angle of incidence always equals the angle of reflection implies that rays R and I can be interchanged.

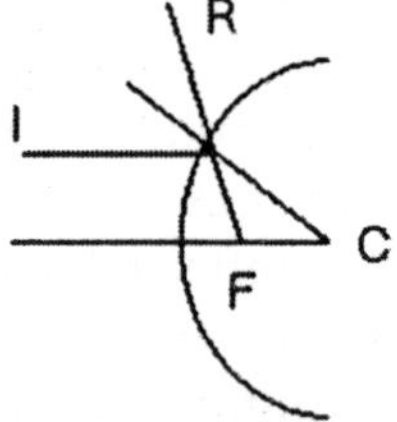

The following diagram shows the same rules applied to a concave mirror, that is one whose curvature is inward away from the source of the rays. A ray that travels parallel to the axis is reflected through the focal point the fact that the angle of incidence always equals the angle of reflection implies that this ray can also be reversed.

That is, a ray striking the mirror from the focal point is reflected parallel to the axis. As above, a ray that strikes the mirror along the radius of curvature is simply reflected back along itself, since any such ray strikes the mirror with an angle of incidence of exactly 0.

The other two rules: a ray directed along the centre of curvature is simply reflected back on itself, since the angle of incidence and reflection are both 0.

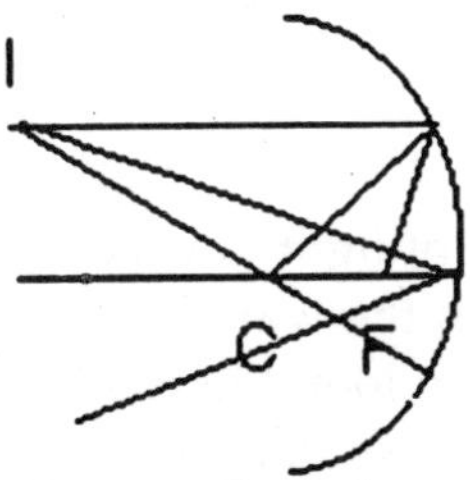

Likewise, a ray that strikes the mirror exactly on its symmetry axis is reflected below the axis by the same angle.

Spherical Mirrors, Image Formation.

The rules of the preceding can be used to calculate the position and size of the image produced by any spherical mirror.

1. The image produced by a convex mirror. The image is formed behind the mirror and is therefore virtual. The image is erect and is smaller than the object. The image is formed at a point behind the mirror that is almost always less than the distance from the front of the mirror to the object. The image distance varies in the same direction as the object distance – an increase in one produces a corresponding increase in the other. However, the maximum image distance is the focal distance when the object is very, very far away, and the minimum image distance is 0 when the object is right at the mirror.

The magnification is the ratio of the size of the image to

the size of the object. Since the object and image vectors form the bases of two similar triangles, the magnification is the ratio of the distance of the image to the focal point divided by the distance of the object to the focal point, and this ratio is almost always less than 1. It approaches 1 as the object comes closer and closer to the mirror surface.

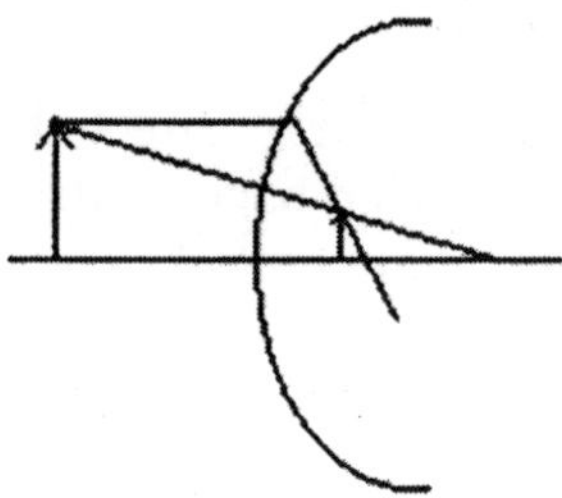

If a point source of light is very far away from a convex mirror, then, although the wavefronts are spheres, they have very large radii and so look like plane waves. The rays that strike the mirror are essentially parallel to each other in this case.

If the rays strike the mirror parallel to its axis, then all of them are reflected so that the reflected rays appear to come from the focal point. The reflected rays therefore look like a single point source of light located at the focal point.

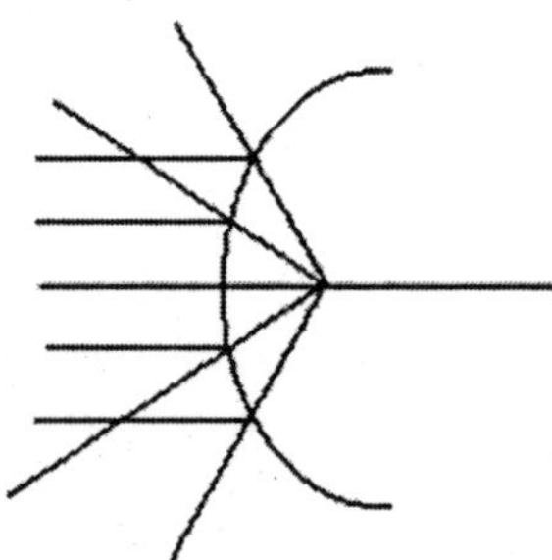

If the rays strike the mirror at an angle to the axis, then image point can be found using the usual 3 rays: one that is aimed at the centre of curvature and is reflected back on itself, one that is aimed at the focal point and is reflected parallel to the axis, and one that is directed at the centre of the mirror and is reflected symmetrically below the axis.

The image point is on the focal plane, but not at the focal point. That is, it is located directly above (or below) the focal point.+

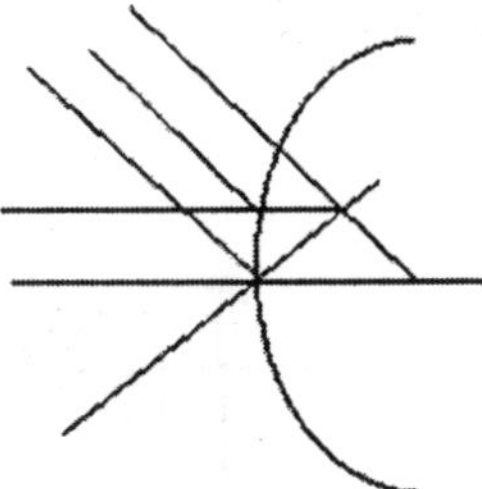

2. The image produced by a concave mirror when the object is outside of the focal point. The image is inverted and is real. The rays intersect at the image point and really do appear to diverge again from that point.

When the object is further away from the mirror than the centre of curvature the image will be formed somewhere between the centre of curvature and the focal length. The image will always be inverted and will be smaller than the object.

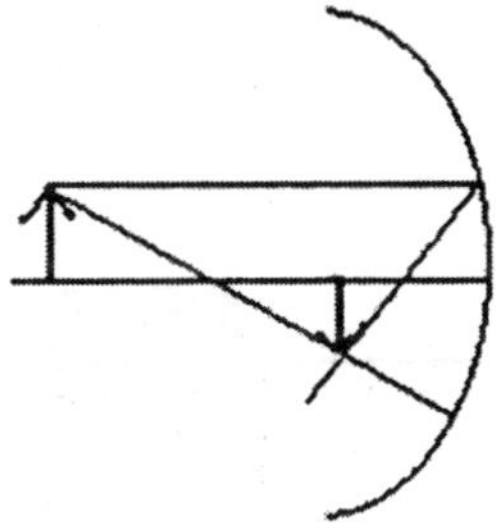

3. As the object in the diagram above moves in towards the mirror, the image moves out to meet it. The following diagram shows the object and image when the object has moved in far enough so that it is located at the centre of curvature of the mirror. No ray from the top of the object can pass through the centre of curvature in this configuration, but we can find the position of the image using other standard rays: a ray parallel to the axis and a ray that strikes the mirror exactly at its centre.

The diagram below is symmetric between the object and the image: both are located at the centre of curvature and are equal in size. The image is inverted, but it has now grown to be the same size as the object.

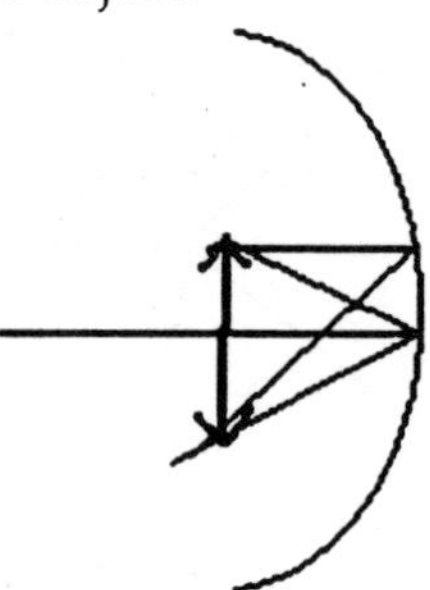

3. This process continues as the object is moved closer and closer to the mirror. The image continues to increase in size, but it remains inverted and it continues to move further and further away from the mirror. Nothing special happens as long as the object does not actually get to the focal point. The image is always inverted, and it is always larger than the object.

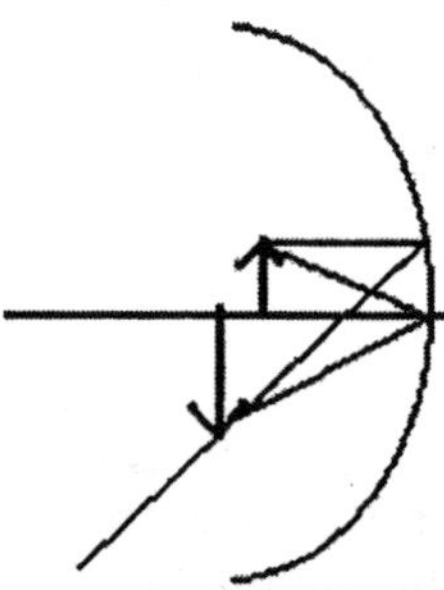

5. The image produced by a concave mirror when the object is at the focal point. The rays from the bottom of the arrow all emerge from the focal point and therefore are reflected parallel to the axis. The rays do not converge to an image point at any finite distance. We can think of this as if the image is formed at infinity, where all of the parallel lines meet again.

Rays from the top of the arrow cannot pass through the centre of curvature or the focal point. A ray that left the top of the arrow parallel to the axis would be reflected through the focal point, and a ray from the top of the arrow that was headed for the centre of the mirror would be reflected downward by the same angle.

These two rays are divergent and do not converge to a single point at any finite distance. This confirms the statement that there is no image of the entire object at any finite distance – only at infinity.These two cases are symmetrical rays entering parallel to the axis from infinity converge at the focal point, and rays emerging from the focal point in any direction are reflected parallel to the axis.

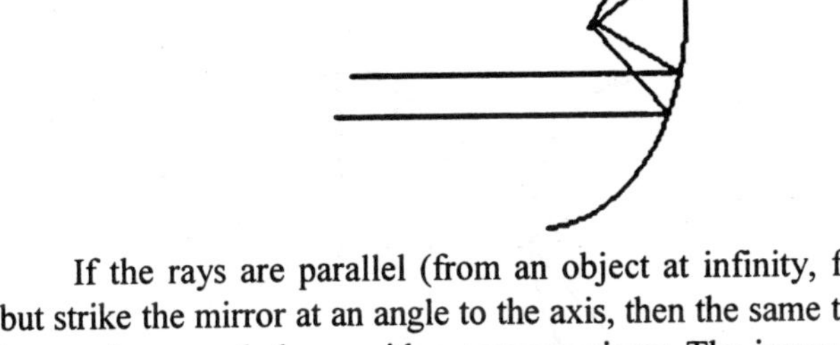

If the rays are parallel (from an object at infinity, for example), but strike the mirror at an angle to the axis, then the same thing happens here as happened above with a convex mirror. The image is formed in the focal plane, but not at the focal point. The position of the image is found using the same technique as was used above.

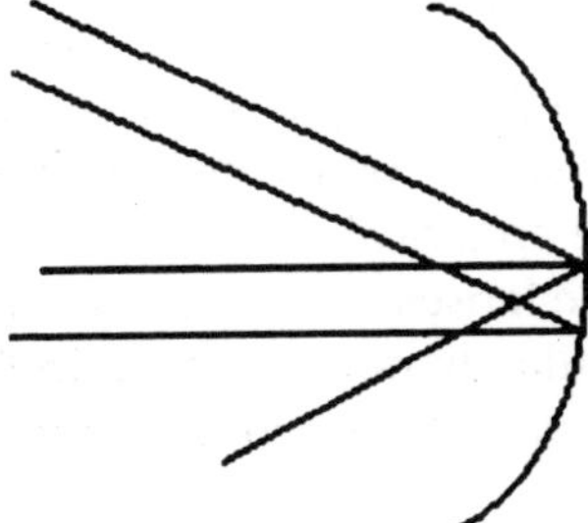

6. When the object moves inside of the focal point, then the image changes in a fundamental way. It is no

longer possible for a ray from the object to actually pass through either the centre of curvature or the focal point, since the object is inside of both of these distances. However, there are still rays which leave the object along these same directions so that they might have seemed to emerge from the focal point or the centre of curvature, and these rays can be used to construct the image in this case. Likewise, a ray which leaves the object headed for the centre of the mirror is reflected symmetrically about the axis of the mirror as usual.

These rays do not converge to any finite point, and therefore there is no real image formed. However, the rays appear to be coming from a point on the right hand side of the mirror, and therefore there is a virtual image formed on the right hand side. Note that this image is not inverted as the real images above were, but is erect. Also, it is larger than the object.

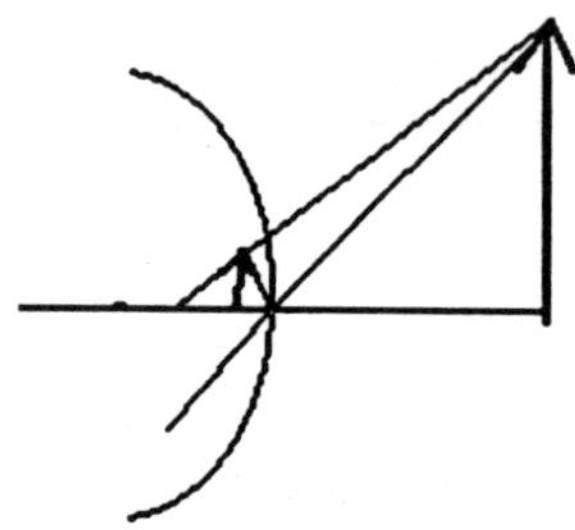

THIN LENSES, TYPES OF LENSES

In addition to using mirrors to form images, it is also possible to form images using lenses. Although many different types of lenses exist, we will consider in detail only lenses whose surfaces are spherical.

In addition, the lenses we will study are made from glass or some other similar material whose index of refraction is greater than the index of refraction of the air that surrounds them..

The lenses we will study fall into two broad categories: converging, or positive lenses and diverging, or negative ones.

Each category has a number of different configurations, but all of the shapes in either category share a common characteristic. Converging, or positive lenses are always thicker in the middle than they are at the edges, while diverging, or negative lenses are always thinner in the middle than they are at the edges.

These are some examples of converging, or positive lenses:

The lens on the left is called a positive meniscus (or "fish-eye") lens, the middle one is a plano-convex lens, and the one on the right is a double-convex lens. Although all of these shapes have the same basic optical properties, the first two produce better images when the field of view is very large or when the object is very far away compared to the size of the lens. The ray-tracing examples will always use the symmetric configuration on the right, but the same ray-tracing rules will work for the other two shapes as well.

Here are some examples of diverging, or negative lenses:

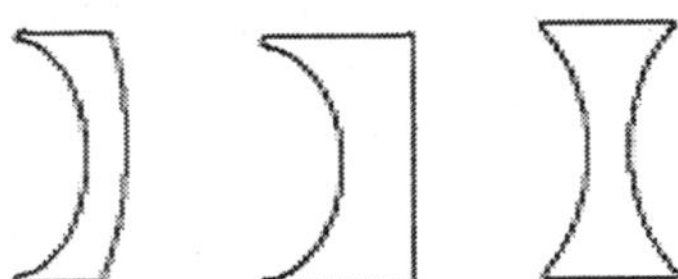

The lens on the left is a negative meniscus lens, the one in the middle is a plano-concave shape and the one on the right is a double-concave shape. The choice among these shapes is governed by the same considerations as for a positive lens as described above. We will use the symmetric concave shape in the ray-tracing examples.

In each case, the properties of the lens are completely specified by its focal length. However, unlike the case of a mirror where the focal length is a simple geometrical parameter, the focal length of a lens depends both on its shape and on the index of refraction of the material it is made of. For the symmetric convex and concave shapes shown above, the approximate focal length is determined from the equation:

$$\frac{1}{f}-\frac{2(n-1)}{R}$$

where R is the radius of curvature of either surface and n is the index of refraction of the glass (or whatever material the lens is made of). The magnitude of the focal length is determined by this relationship for either a double convex or a double concave lens; the sign of the focal length is positive for a converging (convex) lens and negative for a diverging (concave) shape.

Since all lenses have light rays traveling through the lens, all lenses have two focal points: one on each side of the lens. The distance from either focal point to the centre of the lens is given by the expression above.

It is very common for relationships involving the focal length to depend not on the focal length itself but on its reciprocal. This reciprocal appears so often in optical design that it is called the "strength" or "power" of the lens and it is measured in *diopters*.

The strength of a lens in diopters is the reciprocal of its focal length in *meters*. A lens with a focal length of 50 cm (0.5 m) would have a strength of 2 diopters, a focal length of 25 cm would be a strength 4 diopters, etc. Note that the focal length must be converted to meters to compute the strength of the lens.

The sign of the strength of the lens in diopters is the same as the sign of the focal length in meters: positive for convex or converging lenses and negative for diverging ones.

Thin Lenses, Ray Tracing Rules

The ray-tracing rules can be used to find the position and size of an image given the focal length of the lens and the position of the object. The rules depend on the same approximations as for mirrors: the rules are valid only when the rays strike the lens in the paraxial region (i.e., close to the central axis of symmetry).

In addition, the lens must be "thin," which means that its thickness must be small compared to its focal length. Many real-world lenses (such as those in cameras) do not satisfy this condition exactly, but the general principles of this section are still applicable.

Ray-tracing rules for a converging (positive) lens:

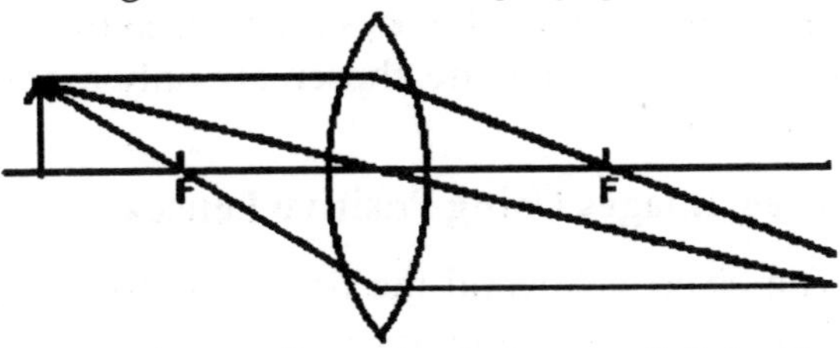

1. A ray that strikes the lens parallel to the axis is refracted through the lens and passes through the focal point on the far side of the lens. (This is often called the second focal point)..
2. A ray that passes through the centre of the lens emerges traveling in the same direction with no change.
3. A ray which passes through the first focal point (on the object side of the lens) is refracted so that it emerges on the other side parallel to the axis of the lens.

Ray tracing rules for a diverging (negative) lens:

1. A ray that strikes the lens parallel to the axis is refracted through the lens so that it appears to come from the focal point on the object side of the lens.

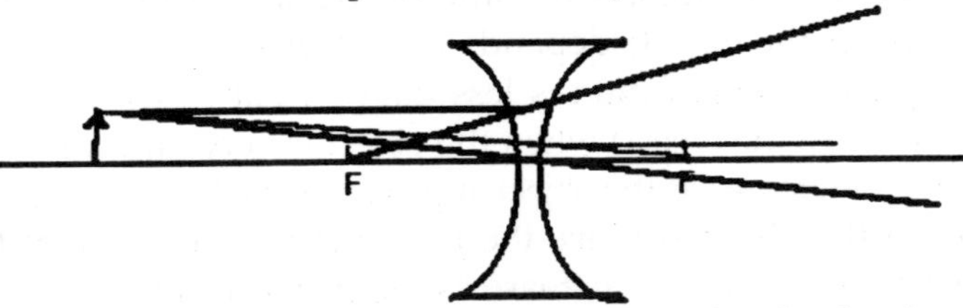

2. A ray that strikes the lens heading for the focal point on the image side of the lens is refracted through the lens so that it emerges parallel to the axis on the other side.
3. A ray that strikes the lens in the centre emerges traveling in the same direction on the image side.

In general, only two of the three rules are needed to locate the image, and the most convenient pair of rules can be chosen in any situation.

The fundamental assumption of these rules is that *all* of the rays from any point on the object will reach the same point of the image: the rays that are chosen are only examples that are easier to draw.

Thin Lenses, Images Using Positive Lenses

1. The object is beyond the first focal point, but is at a finite distance from the lens:

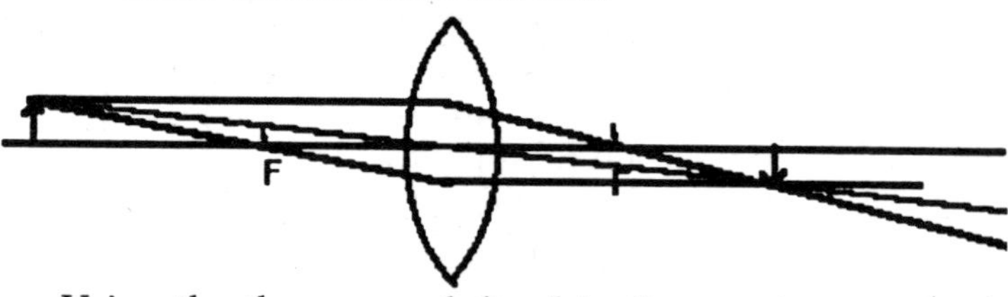

Using the three rays defined in the previous topic, the image is real, inverted and beyond the second focal point on the image side of the lens.

The image is real because the rays on the image side of the lens from any point on the object really do converge to a single point on the image. If we look into the lens from the image side the rays appearing to diverge from the image that is formed.

The magnification (the ratio of the size of the image to the size of the object) depends on the exact position of the object. The magnification is less than 1 (that is, the image is smaller than the object) when the object is very far away, and it slowly increases as the object approaches the lens. It is exactly 1 when the object is twice the focal distance away from the lens, and continues to increase as the distance is reduced still further. The lens produces a magnified image when the object is located at a distance of less than twice the focal length (but greater than the focal length) from the lens.

Note that in addition to getting larger, the image moves away from the lens (to the right) as the object approaches the lens.

2. The object is at the first focal point on the left side of the lens.

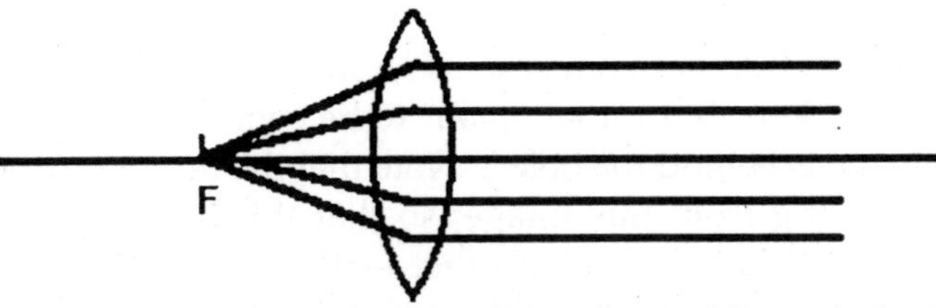

Since every ray leaving the object passes through the focal point, all of the rays emerge parallel to the axis on the right-hand side of the lens.

A point object located at this position therefore produces a parallel beam of light.

Note that this configuration can be used to show that a beam of parallel light striking the lens is imaged to a point of light at the focal point.

This configuration is simply the inverse of the situation that is shown and would be realised by simply interchanging the object and the image.

This configuration can be used to determine the focal length of a positive lens – it is simply the location of the image when the object is very far away. As a practical matter, an object distance of more than 10 times the focal length is sufficient for most purposes.

3. The object is inside of the first focal point on the left side of the lens.

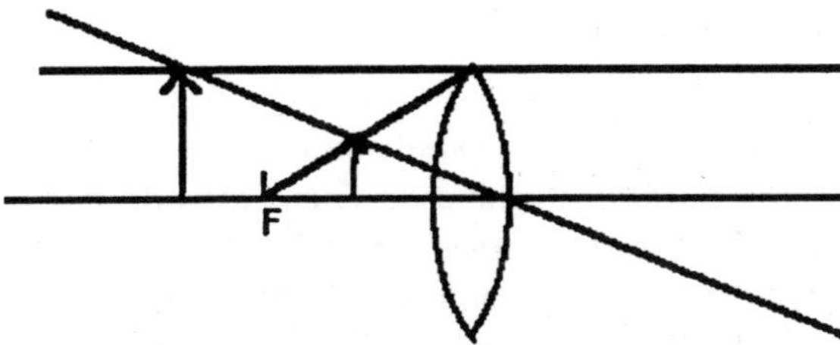

The blue ray that leaves the object comes off at the same angle as a ray that originated at the focal point, which is to the left of the object in this situation. Therefore, this blue ray is bent so that it emerges parallel to the axis on the right side of the lens.

The green ray that leaves the object heading towards the centre of the lens passes through the lens without any change in direction.

The rays do not converge to an image on the right hand side of the lens, but rather *appear* to be coming from a larger image that is behind the object. Note that the rays only appear to be coming from this image, so that this image is virtual rather than real. It is also erect – that is it has the same orientation as the object and it is magnified – its size is larger than the size of the object.

Thin Lenses, Images using Negative Lenses

The situation when a negative lens is used to form an image is simpler than for a positive lens, because the general features of the image are the same no matter where the object is located.

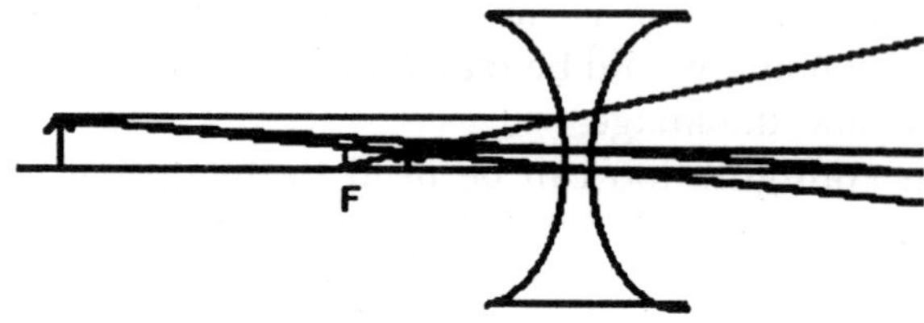

The green ray, which strikes the lens parallel to the axis emerges on the other side of the lens as if it had originated at the focal point a shown.

The blue ray, which strikes the lens at its centre, emerges on the other side without any change in its direction.The purple ray, which strikes the lens in a direction that is headed for the focal point on the opposite side of the lens, emerges on the other side parallel to the axis.These rays do not converge to an image. Rather, they appear to come from a virtual image which is smaller than the object, has the same orientation as the object, and is located between the lens and the focal t figure.

FRESNEL LENSES

A point source of light at the focal point of a positive lens produces an image which is a parallel beam of light. This configuration is used in spotlights, lighthouses, and similar applications. In order to minimize the size of the device, the focal length of the lens must be as short as possible, since this sets the distance between the lens and the lamp. This short focal length is usually implemented in a very large lens so as to capture as much of the output of the lamp as possible. The resulting lens is large and very heavy.

Since the imaging power of a lens is produced by refraction at its edges, the centre of the lens makes almost no contribution. A Fresnel lens is designed using this idea. The interior glass of the lens is removed, and the curved surface of the lens is approximated by a series of curved segments. The effect is to preserve most of the focusing power of the lens while at the same time saving a considerable amount on its size and weight.

The design of a Fresnel lens usually starts from a plano-convex lens in the left panel. The lens is then cut into horizontal sections and each section is then aligned so that the curved edges on the right hand side of the lens are aligned. Then the extra material on the back side of the lens is removed.

The resulting lens has the same refractive performance as the original, since the shapes of both the front and the rear surfaces have not changed. However, it has much less material and is therefore much lighter.

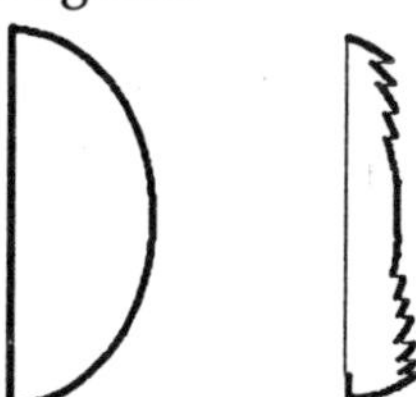

Although it looks like the modified Fresnel should have the same performance as the original, this is not quite true. A ray through the centre of the Fresnel lens, for example, reaches a point on the image side much faster than a ray through the centre of the original lens, because a larger fraction of its path is in air. Thus a packet of rays which travel through the Fresnel lens are separated in time compared to the same packet of rays passing through the original lens. This same effect is true for rays that travel through other parts of the lens. This results in subtle changes to the image that are usually not important for spotlights and lighthouses, but can be significant whenever the phase of the light is important.

To build a spotlight or a lighthouse, a Fresnel lens is usually combined with a concave mirror. The concave mirror is placed behind the lamp with the lamp at its centre of curvature. Every ray that leaves the lamp traveling backwards strikes the mirror perpendicular to its surface in this case and is therefore reflected back on itself. These reflected rays therefore travel in the same directions as the rays that emerge from the front of the lamp.

The lamp is placed at the focal point of the Fresnel lens, so that all of these rays emerge from the lens in a parallel beam. This configuration uses most of the light that is emitted by the lamp, although the rays that do not strike either the mirror or the lens are wasted.

Chapter 4

Wave Optics

Geometric optics is an incredibly successful theory. Probably its most important application is in describing and explaining the operation of commonly occurring optical instruments: *e.g.*, the camera, the telescope, and the microscope. Although geometric optics does not make any explicit assumption about the nature of light, it tends to suggest that light consists of a stream of massless particles. This is certainly what scientists, including, most notably, Isaac Newton, generally assumed up until about the year 1800.

Let us examine how the particle theory of light accounts for the three basic laws of geometric optics:

- *The law of geometric propagation:* This is easy. Massless particles obviously move in straight-lines in free space.
- *The law of reflection:* This is also fairly easy. We merely have to assume that light particles bounce *elastically* (*i.e.*, without energy loss) off reflecting surfaces.
- *The law of refraction:* This is the tricky one. Let us assume that the speed of light particles propagating through a transparent dielectric medium is proportional to the index of refraction, $n \equiv \sqrt{K}$. Let us further assume that at a general interface between two different dielectric media, light particles crossing the interface conserve momentum in the plane parallel to the interface. In general, this implies that the particle momenta normal to the interface are not conserved: *i.e.*, the interface exerts a normal reaction force on crossing particles, but no parallel force.

Parallel momentum conservation for light particles crossing the interface yields $v_1 \sin \theta_1 = v_2 \sin \theta_2$.

- However, by assumption, $v_1 = n_1 c$ and $v_2 = n_2 c$, so

$$n_1 \sin\theta_1 = n_1 \sin\theta_2.$$

- This highly contrived (and incorrect) derivation of the law of refraction was first proposed by Descartes in 1637. Note that it depends crucially on the (incorrect) assumption that light travels *faster* in dense media (*e.g.*, glass) than in rarefied media (*e.g.*, water). This assumption appears very strange to us nowadays, but it seemed eminently reasonable to scientists in the 17th and 18th centuries. After all, they knew that sound travels faster in dense media (*e.g.*, water) than in rarefied media (*e.g.*, air).

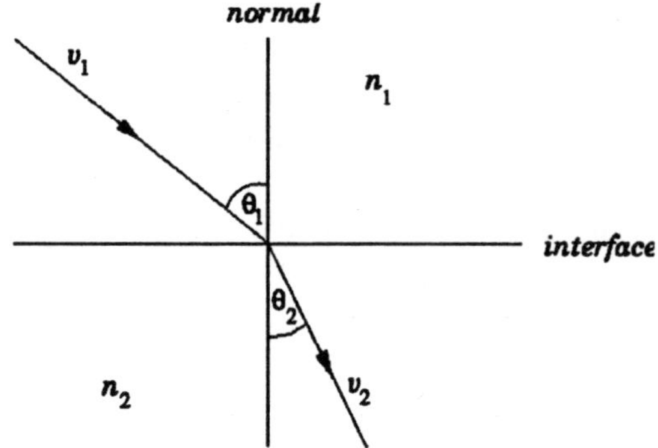

Fig. Descartes' Model of Refraction

The wave theory of light, which became established in the first half of the 19th century, initially encountered tremendous resistance. Let us briefly examine the reasons why scientists in the early 1800s refused to think of light as a wave phenomenon? Firstly, the particle theory of light was intimately associated with Isaac Newton, so any attack on this theory was considered to be a slight to his memory. Secondly, all of the waves that scientists were familiar with at that time manifestly did not travel in straight-lines. For instance, water waves are diffracted as they pass through the narrow mouth of a harbour. In other words, the "rays" associated with such waves are bent as they traverse the harbour mouth. Scientists thought that if light were a wave phenomenon then it would also not travel in straight-lines: *i.e.*, it would not cast straight, sharp shadows, any more than water waves cast straight, sharp "shadows." Unfortunately, they did not appreciate that if the wavelength of light is much shorter than that of water waves then light can be a wave phenomenon and still propagate in a largely geometric manner.

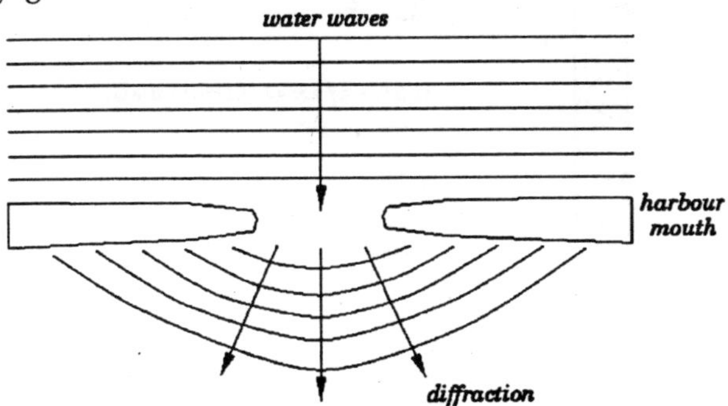

Fig. Refraction of Water Waves Through the Entrance of a Harbour.

HUYGENS' PRINCIPLE OF WAVE

The first person to explain how wave theory can also account for the laws of geometric optics was Christiaan Huygens in 1670. At the time, of course, nobody took the slightest notice of him. His work was later rediscovered after the eventual triumph of wave theory.

Huygens had a very important insight into the nature of wave propagation which is nowadays called *Huygens' principle.* When applied to the propagation of light waves, this principle states that:

Every point on a wave-front may be considered a source of secondary spherical wavelets which spread out in the forward direction at the speed of light. The new wave-front is the tangential surface to all of these secondary wavelets.

According to Huygens' principle, a plane light wave propagates though free space at the speed of light, c. The light rays associated with this wave-front propagate in straight-lines. It is also fairly straightforward to account for the laws of reflection and refraction using Huygens' principle.

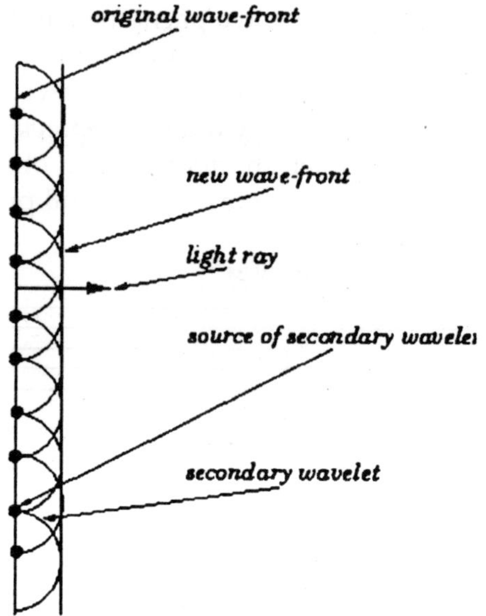

Fig. Huygen's Principle.

YOUNG'S DOUBLE-SLIT EXPERIMENT

The first serious challenge to the particle theory of light was made by the English scientist Thomas Young in 1803. Young possessed one of the most brilliant minds in the history of science. A physician by training, he was the first to describe how the lens of the human eye changes shape in order to focus on objects at differing distances. He also studied Physics, and, amongst other things, definitely established the wave theory of light, as described below. Finally, he also studied Egyptology, and helped decipher the Rosetta Stone.

Young knew that sound was a wave phenomenon, and, hence, that if two sound waves of equal intensity, but 180°out of phase, reach the ear then they cancel one another out, and no sound is heard. This phenomenon is called *interference.* Young reasoned that if light were actually a wave phenomenon, as he suspected, then a similar interference effect should occur for light. This line of reasoning lead Young to perform an experiment which is nowadays referred to as *Young's double-slit experiment.*

In Young's experiment, two very narrow parallel slits, separated by a distance *d*, are cut into a thin sheet of metal. Monochromatic light, from a distant light-source, passes through the slits and eventually hits a screen a comparatively large distance *L* from the slits.

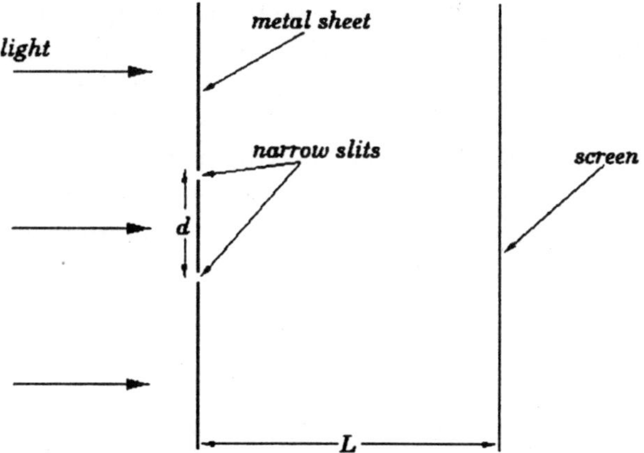

Fig. Young's Double–Slit Experiment.

According to Huygens' principle, each slit radiates spherical light waves. The light waves emanating from each slit are superposed on the screen. If the waves are 180°out of phase then *destructive interference* occurs, resulting in a dark patch on the screen.

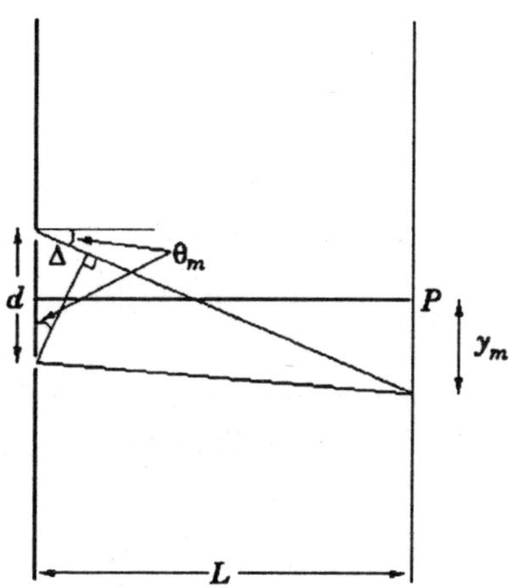

Fig. Interference of Light in Young's Double–Slit Experiment.

On the other hand, if the waves are completely in phase then *constructive interference* occurs, resulting in a light patch on the screen.

The point P on the screen which lies exactly opposite to the centre point of the two slits is obviously associated with a bright patch. This follows because the path-lengths from each slit to this point are the same. The waves emanating from each slit are initially in phase, since all points on the incident wave-front are in phase (*i.e.*, the wave-front is straight and parallel to the metal sheet). The waves are still in phase at point P since they have traveled equal distances in order to reach that point.

The general condition for constructive interference on the screen is simply that the difference in path-length Δ between the two waves be an *integer* number of wavelengths. In other words,

$$\Delta = m\lambda,$$

where m=0, 1, 2,. . . . Of course, the point P corresponds to the special case where m= 0. It follows that the angular location of the m th bright patch on the screen is given by

$$\sin\theta_m = \frac{\Delta}{d} = \frac{m\lambda}{d}.$$

Likewise, the general condition for destructive interference on the screen is that the difference in path-length between the two waves be a *half-integer* number of wavelengths. In other words,

$$\Delta = (m + 1/2)\lambda,$$

where m = 1, 2, 3, · · ·. It follows that the angular coordinate of the mth dark patch on the screen is given by

$$\sin\theta'_m = \frac{\Delta}{d} = \frac{(m + 1/2)\lambda}{d}.$$

Usually, we expect the wavelength λ of the incident light to be much less than the perpendicular distance L to the screen. Thus,

$$\sin\theta_m \simeq \frac{y_m}{L},$$

where y_m measures position on the screen relative to the point P.

It is clear that the interference pattern on the screen consists of *alternating light and dark bands,* running parallel to the slits. The distances of the centers of the various light bands from the point P are given by

$$y_m = \frac{m\lambda L}{d},$$

where $m = 0,1,2, \cdots$.. Likewise, the distances of the centres of the various dark bands from the point P are given by

$$y_m' = \frac{(m+1/2)\lambda L}{d},$$

where $m = 1, 2, 3, \cdots$. The bands are *equally spaced,* and of thickness $\lambda L/d$. Note that if the distance from the screen L is much larger than the spacing d between the two slits then the thickness of the bands on the screen greatly exceeds the wavelength λ of the light. Thus, given a sufficiently large ratio L/d, it should be possible to observe a banded interference pattern on the screen, despite the fact that the wavelength of visible light is only of order 1 micron. Indeed, when Young performed this experiment in 1803 he observed an interference pattern of the type described above. Of course, this pattern is a *direct consequence* of the wave nature of light, and is completely inexplicable on the basis of geometric optics.

It is interesting to note that when Young first presented his findings to the Royal Society of London he was ridiculed. His work only achieved widespread acceptance when it was confirmed, and greatly extended, by the French physicists Augustin Fresnel and Francois Argo in the 1820s. The particle theory of light was dealt its final death-blow in 1849 when the French physicists Fizeau and Foucault independently demonstrated that light propagates *more slowly* though water than though air.

Recall that the particle theory of light can only account for the law of refraction on the assumption that light propagates *faster* through dense media, such as water, than through rarefied media, such as air.

Interference in Thin Films

In everyday life, the interference of light most commonly gives rise to easily observable effects when light impinges on a thin film of some transparent material. For instance, the brilliant colours seen in soap bubbles, in oil films floating on puddles of water, and in the feathers of a peacock's tail, are due to interference of this type.

Suppose that a very thin film of air is trapped between two pieces of glass. If monochromatic light (*e.g.*, the yellow light from a sodium lamp) is incident *almost normally* to the film then some of the light is reflected from the interface between the bottom of the upper plate and the air, and some is reflected from the interface between the air and the top of the lower plate. The eye focuses these two parallel light beams at one spot on the retina. The two beams produce either destructive or constructive interference, depending on whether their path difference is equal to an odd or an even number of half-wavelengths, respectively.

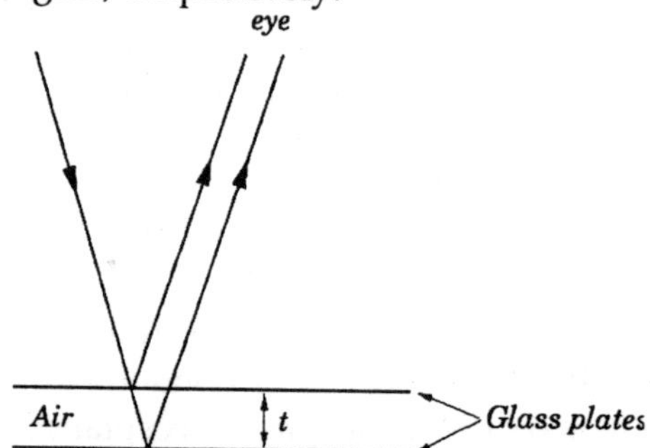

Fig. Interference of light Due to a Thin Film of air Trapped between two Pieces of Glass.

Let t be the thickness of the air film. The difference in path-lengths between the two light rays shown in the figure is clearly $\Delta = 2t$. Naively, we might expect that constructive interference, and, hence, *brightness*, would occur if $\Delta = m\lambda$, where m is an integer, and destructive interference, and, hence, *darkness*, would occur if $\Delta = (m + 1/2)\lambda$. However, this is not the entire picture, since an additional phase difference is introduced between the two rays on reflection. The first ray is reflected at an interface between an optically dense medium (glass), through which the ray travels, and a less dense medium (air).

There is no phase change on reflection from such an interface, just as there is no phase change when a wave on a string is reflected from a free end of the string. (Both waves on strings and electromagnetic waves are *transverse waves*, and, therefore, have analogous properties.) The second ray is reflected at an interface between an optically less dense medium (air), through which the ray travels, and a dense medium (glass). There is a 180° phase change on reflection from such an interface, just as there is a 180° phase change when a wave on a string is reflected from a fixed end.

Thus, an additional 180° phase change is introduced between the two rays, which is equivalent to an additional path difference of $\lambda/2$. When this additional phase change is taken into account, the condition for constructive interference becomes where *m* is an integer. Similarly, the condition for destructive interference becomes

$$2t = m\lambda.$$

For white light, the above criteria yield constructive interference for some wavelengths, and destructive interference for others. Thus, the light reflected back from the film exhibits those colours for which the constructive interference occurs.

If the thin film consists of water, oil, or some other transparent material of refractive index *n* then the results are basically the same as those for an air film, except that the wavelength of the light in the film is reduced from λ (the vacuum wavelength) to λ/n. It follows that the modified criteria for constructive and destructive interference are

$$2nt = (m + 1/2)\lambda, \text{ and}$$

$2nt = m\lambda$, respectively.

SIMPLE HARMONIC MOTION

Simple harmonic motion is the motion of a simple harmonic oscillator, a motion that is neither driven nor damped. The motion is periodic, as it repeats itself at standard intervals in a specific manner - described as being sinusoidal, with constant amplitude.

It is characterized by its amplitude (which is always positive), its period which is the time for a single oscillation, its frequency which is the number of cycles per unit time, and its phase, which determines the starting point on the sine wave. The period, and its inverse the frequency, are constants determined by the overall system, while the amplitude and phase are determined by the initial conditions (position and velocity) of that system.

Simple harmonic motion is defined by the differential equation $m\frac{d^2x}{dt^2} = -kx$, where "k" is a positive constant, "m" is the mass of the body, and "x" is its displacement from the mean position.

In words, simple harmonic motion is "motion where the force acting on a body and thereby acceleration of the body is proportional to, and opposite in direction to the displacement from its equilibrium position" (i.e. $F =$ " kx).

A general equation describing simple harmonic motion is $x(t) = A \cos(2\pi ft + \phi)$, where x is the displacement, A is the amplitude of oscillation, f is the frequency, t is the elapsed time, and ö is the phase of oscillation. If there is no displacement at time t = 0, the phase $\phi = \frac{\pi}{2}$. A motion with frequency f has period, $T = \frac{1}{f}$.

Simple harmonic motion can serve as a mathematical model of a variety of motions and provides the basis of the characterization of more complicated motions through the techniques of Fourier analysis.

MATHEMATICS OF HARMONIC MOTION

It can be shown, by differentiating, exactly how the acceleration varies with time. Using the angular frequency ù, defined as

$$\text{w} = 2\pi f,$$

the displacement is given by the function

$$x(t) = A \cos(\omega t + f).$$

Differentiating once gives an expression for the velocity at any time.

$$v(t) = \frac{d\,x(t)}{dt} = -A\omega \sin(\omega t + \phi).$$

And once again to get the acceleration at a given time.

$$a(t) = \frac{d^2\,x(t)}{dt^2} = -A\omega^2 \cos(\omega t + \phi).$$

These results can of course be simplified, giving us an expression for acceleration in terms of displacement.

$$a(t) = -\,\omega^2 x(t).$$

$$a(t) = -(2\pi f)^2 x(t).$$

When and if total energy is constant and kinetic, the formula $E = \frac{kA^2}{2}$ applies for simple harmonic motion, where E is considered the total energy while all energy is in its kinetic form. A representing the mean displacement of the spring from its rest position in MKS units.

Mass on a Spring

A mass M attached to a spring of spring constant k exhibits simple harmonic motion in space with

$$\omega = 2\pi f = \sqrt{\frac{k}{M}}.$$

Alternately, if the other factors are known and the period is to be found, this equation can be used:

$$T = \frac{1}{f} = 2\pi\sqrt{\frac{M}{k}}.$$

The total energy is constant, and given by $E = \frac{kA^2}{2}$,where E is the total energy.

Uniform Circular Motion

Simple harmonic motion can in some cases be considered to be the one-dimensional projection of uniform circular motion.

If an object moves with angular frequency ω around a circle of radius *R* centered at the origin of the x-y plane, then its motion along the x and the y coordinates is simple harmonic with amplitude *R* and angular speed ω.

Mass on a Pendulum

In the small-angle approximation, the motion of a pendulum is approximated by simple harmonic motion. The period of a mass attached to a string of length ℓ with gravitational acceleration g is given by

$$T = 2\pi\sqrt{\frac{\ell}{g}}.$$

This approximation is accurate only in small angles because of the expression for angular acceleration being proportional to the sine of position:

$$\ell mg \sin(\theta) = I\alpha$$

where I is the moment of inertia; in this case $I = m\ell^2$. When \ is small, sin (θ) ≈ θ and therefore the expression becomes

$$\ell mg(\theta) = I\alpha$$

which makes angular acceleration directly proportional to \, satisfying the definition of Simple Harmonic Motion.

For a solution not relying on a small-angle approximation (mathematics).

USEFUL FORMULAE

Given mass *M* attached to a spring/pendulum with amplitude *A* with acceleration *a*:

$$k = \frac{Ma}{A}$$

$$f = \frac{A}{t} = \frac{\lambda}{t}$$

$$T_s = T_p = \frac{1}{f} = \frac{t}{A} = 2\pi\sqrt{\frac{M}{k}} = 2\pi\sqrt{\frac{A}{g}} = 2\pi\sqrt{\frac{\ell}{g}}.$$

$$E_{tot} = \frac{kA^2}{2} = \frac{MaA}{2}.$$

Where:

k is the spring constant.
M is the mass (usually in kilograms)
a is the acceleration.
A is the amplitude OR ë is the wavelength.
f is the frequency (usually in hertz).
t is the time in seconds to complete one cycle.
T_s or T_p is the period of the spring or pendulum.
g is the acceleration due to gravity (On Earth at sea level: 9.81 m/s²).
is the length of the pendulum.
E_{tot} is the total energy.

THE SPRING PENDULUM

Another mecahnical example will serve to establish the paradigm of SIMPLE HARMONIC MOTION (*SHM*) as a solution to a particular type of *equation of motion.*

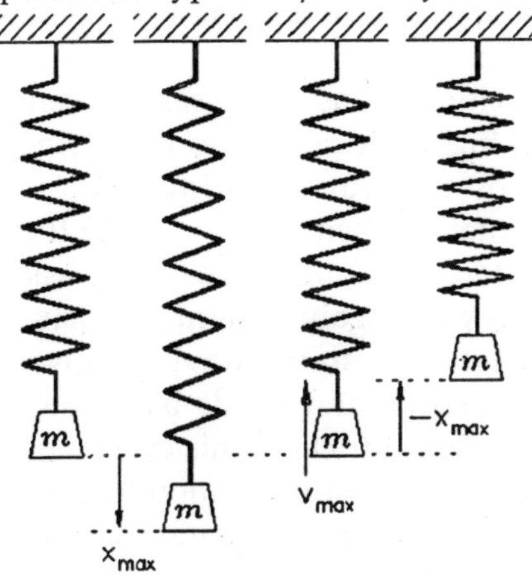

Fig. Successive "Snapshots" of a Mass Bouncing up and Down on a Spring.

The *spring* embodies one of Physics' premiere paradigms, the *linear restoring force*. That is, a force which disappears when the system in question is in its "equilibrium position" x_0 [which we will define as the $x = 0$ position $(x_0 \equiv 0)$ to make the calculations easier] but increases as x moves away from equilibrium, in such a way that the *magnitude* of the force F is proportional to the displacement from equilibrium [F is *linear* in x] and the *direction* of F is such as to try to *restore* x to the original position. The constant of proportionality is called the *spring constant*, always written k. Thus $F = -k\,x$ and the resultant equation of motion is

$$\ddot{x} = -\left(\frac{k}{m}\right)x.$$

linear restoring force in this paradigm. If $m \to 0$ in this equation, then the acceleration becomes infinite and in principle the spring would just return instantaneously to its equilibrium length and stay there!

The mass m is at rest and the spring is in its equilibrium position (*i.e.* neither stretched nor compressed) defined as x=0. In the second frame, the spring has been gradually pulled down a distance x_{max} and the mass is once again at rest. Then the mass is released and accelerates upward under the influence of the spring until it reaches the equilibrium position again [third frame]. This time, however, it is moving at its maximum velocity v_{max} as it crosses the centre position; as soon as it goes higher, it *compresses* the spring and begins to be *decelerated* by a linear restoring force in the opposite direction. Eventually, when $x = -x_{max}$, all the kinetic energy has been been stored back up in the compression of the spring and the mass is once again instantaneously at rest [fourth frame]. It immediately starts moving downward again at maximum acceleration and heads back toward its starting point. In the absence of friction, this cycle will repeat forever.

I now want to call your attention to the acute similarity between the above differential equation and the one we solved for exponential decay:

$$\dot{x} = -kx$$

and, by extension,

$$\ddot{x} = k^2 t$$

the solution to which equation of motion (*i.e.* the function which "satisfies" the differential equation) was

$$x(t) = -\,x_0 e^{-kt}$$

Now, if only we could equate k^2 with $-k/m$, these equations of motion (and therefore their solutions) would be exactly the same! The problem is, of course, that both k and m are intrinsically positive constants, so it is tough to find a real constant k which gives a negative number when squared.

Imaginary Exponents

Mathematics, of course, provides a simple solution to this problem: just have k be an *imaginary* number, say

$$k \equiv i\,\omega \text{ where } i \equiv \sqrt{-1}$$

and ω is a positive real constant. Let's see if this trial solution "works" (*i.e.* take its second derivative and see if we get back our equation of motion):

$$\begin{aligned} x(t) &= x_0 e^{i\omega t} \\ \dot{x} &= i\,x_0 e^{i\omega t} \\ \ddot{x} &= -\,\omega^2 x_0 e^{i\omega t} \\ \text{or } \ddot{x} &= -\,\omega^2 x \\ \text{so } \omega &\equiv \sqrt{\frac{k}{m}} \end{aligned}$$

OK, it works. But what does it *describe*? For this we go back to our series expansions for the exponential, sine and cosine functions and note that *if we let* $z \equiv i\theta$, the following *mathematical identity* holds:[13.5]

$$e^{i\theta} = \cos(\theta) + i\sin(\theta).$$

Thus, for the case at hand, if $\theta \equiv \omega\,t$ then

$$x_0 e^{i\omega t} = x_0 \cos(\omega t) + i\,x_0 \sin(\omega t)$$

— *i.e.* the formula for the projection of uniform circular motion, with an imaginary part "tacked on." What does *this* mean?

Keep the imaginary components through all your calculations until the final "answer," and then *throw away* any remaining *imaginary parts* of any actual *measurable quantity*.

The point is, there is a difference between understanding how something *works* and knowing what it *means*. Meaning is something we put into our world by act of will, though not always conscious will. How it works is there before us and after we are gone. No one asks the "meaning" of a screwdriver or a carburetor or a copy machine; some of the conceptual tools of Physics are in this class, though of course there is nothing to prevent anyone from *putting* meaning into them.

PERIODIC BEHAVIOUR

Nature shows us many "systems" which return periodically to the same initial state, passing through the same sequence of intermediate states every period. Life is so full of periodic experiences, from night and day to the rise and fall of the tides to the phases of the moon to the annual cycle of the seasons, that we all come well equipped with "common sense" tailored to this paradigm.[13.1] It has even been suggested that the concept of *time* itself is rooted in the *cyclic* phenomena of Nature.

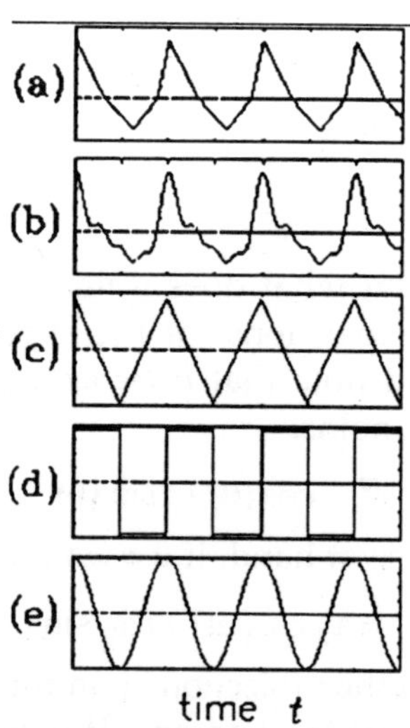

Fig. Some Periodic Functions.

In Physics, of course, we insist on narrowing the definition just enough to allow precision. For instance, many phenomena are *cyclic* without being *periodic* in the strict sense of the word. Here *cyclic* means that the same general pattern keeps repeating; *periodic* means that the system passes through the same "phase" at *exactly* the same time in every cycle and that all the cycles are *exactly* the same length. Thus if we know all the details of *one full cycle* of true periodic behaviour, then we know the subsequent state of the system at *all* times, future and past. Naturally, this is an idealization; but its utility is obvious.

Of course, there is an infinite variety of possible *periodic* cycles. Assuming that we can reduce the "state" of the system to a single variable "q" and its time derivatives, the graph of $q(t)$ can have any shape as long as it *repeats* after one full period.

SINUSOIDAL MOTION

There is one sort of periodic behaviour that is mathematically the *simplest possible* kind. This is the "sinusoidal" motion, so called because one realization is the sine function, sin (x). It is easiest to see this by means of a crude mechanical.

Projecting the Wheel

Imagine a rigid wheel rotating at constant angular velocity about a fixed central axle. A bolt screwed into the rim of the wheel executes *uniform circular motion* about the centre of the axle.[13.3] For reference we scribe a line on the wheel from the centre straight out to the bolt and call this line the *radius vector*. Imagine now that we take this apparatus outside at high noon and watch the motion of the *shadow* of the bolt on the ground. This is (naturally enough) called the *projection* of the circular motion onto the horizontal axis.

At some instant the radius vector makes an angle $\theta = \omega t$ with the horizontal, where ω is the angular frequency of the wheel (2π times the number of full revolutions per unit time) and we measure the time t from the instant when the radius vector is horizontal. From a side view of the wheel we can see that the distance x from the shadow of the central axle to the shadow of the bo : [*i.e.* the *projected horizontal displacement* of the bolt from the centre, where x=0] will be given by trigonometry on the indicated right-angle triangle:

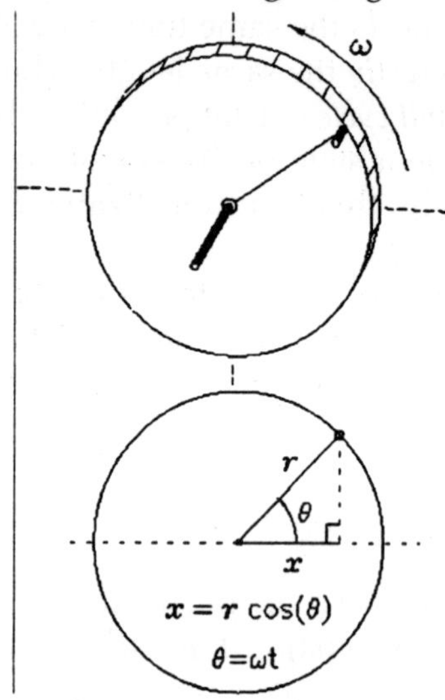

Fig. Projected Motion of a Point on the Rim of a Wheel.

$$\cos(\theta) \equiv \frac{x}{r} \Rightarrow x = r\cos(\theta) = r\cos(\omega t)$$

The resultant *amplitude* of the displacement as a *function of time.*

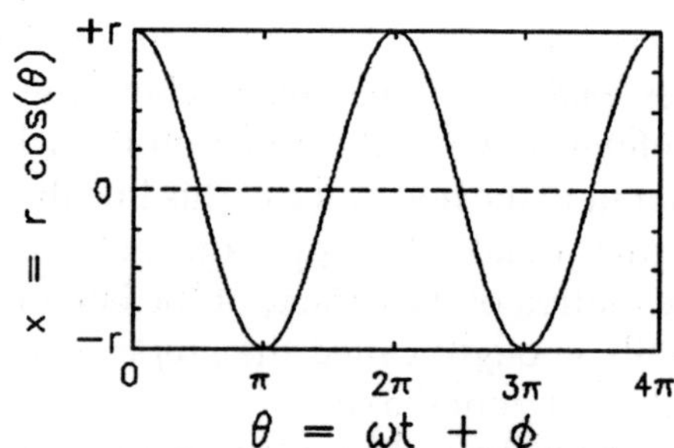

Fig. The Cosine Function.

The *horizontal velocity* v_x of the projected shadow of the bolt on the ground can also be obtained by trigonometry if we recall that the vector velocity $\vec{v}$ is always perpendicular to the radius vector $\vec{r}$.

$$v_x = -v \sin(\theta) = -r\,\omega \sin(\omega t)$$

where $v = r\omega$ is the constant speed of the bolt in its circular motion around the axle. It also follows (by the same sorts of arguments) that the *horizontal acceleration* a_x of the bolt's shadow is the projection onto the $\hat{x}$ direction of $\vec{a}$, which we know is back toward the centre of the wheel — *i.e.* in the $-\hat{x}$ direction; its value at time t is given by

$$a_x = -a\cos(\theta) = -r\omega^2 \cos(\omega t).$$

where $a \frac{v^2}{r} = r\omega^2$ is the magnitude of the centripetal acceleration of the bolt.

DAMPED HARMONIC MOTION

Let's take stock.

$$x(t) = [\text{constant}] - \frac{v_0}{k}\, e^{-kt}$$

satisfies the basic differential equation

$$\ddot{x} = -k\dot{x} \text{ or } a = -kv$$

defining *damped* motion (*e.g.* motion under the influence of a frictional force proportional to the velocity). We now have a solution to the equation of motion defining *SHM*,

$$\ddot{x} = -\omega^2 x \Rightarrow x(t) = x_0\, e^{i\omega t},$$

where

$$\omega = \sqrt{\frac{k}{w}}.$$

Can we put these together to "solve" the more general (and realistic) problem of *damped harmonic motion*? The differential equation would then read

$$\ddot{x} = -\omega^2 x - k\dot{x}$$

which is beginning to look a little hard. Still, we can sort it out: the first term on the *RHS* says that there is a linear restoring force and an inertial factor. The second term says that there is a damping force proportional to the velocity. So the differential equation itself is not all that fearsome. How can we "solve" it?

As always, by trial and error. Since we have found the *exponential* function to be so useful, let's try one here: *Suppose* that

$$x(t) = x_0 e^{Kt}$$

where x_0 and K are unspecified constants. Now plug this into the differential equation and see what we get:

$$\dot{x} = K\, x_0\, e^{Kt} = K\, x$$

and

$$\ddot{x} = K^2\, x_0\, e^{Kt} = K^2\, x$$

The whole thing then reads

$$K^2 x = -\,\omega^2 x - k\, K\, x$$

which can be true "for all x" only if

$$K^2 x = -\,\omega^2 x - kK \text{ or } K^2 + kK + \omega^2 = 0$$

which is in the standard form of a general quadratic equation for K, to which there are two solutions:

$$K = \frac{-k \pm \sqrt{k^2 - 4\omega^2}}{2}$$

Either of the two solutions given by substituting Eq will satisfy describing *damped SHM*. In fact, generally any *linear combination* of the two solutions will also be a solution. This can get complicated, but we *have* found the answer to a rather broad question.

Limiting Cases

Let's consider a couple of "limiting cases" of such solutions. First, suppose that the linear restoring force is extremely weak compared to the "drag" force - *i.e.* $k >> \omega = \sqrt{\frac{k}{m}}$. Then $\sqrt{k^2 - 4\omega^2} \approx k$ and the solutions are

$K \approx -0$ [i.e. $x \approx$ constant, plausible only if $x = 0$] and $K \approx -k$, which gives the same sort of damped behaviour as if there were no restoring force, which is what we expected.

Now consider the case where the linear restoring force is very strong and the "drag" force extremely weak - *i.e.* $k << \omega = \sqrt{\dfrac{k}{m}}$. Then $\sqrt{k^2 - 4\omega^2} \approx 2\,i\omega$ and the solutions are $K \approx -\frac{1}{2}k \pm i\omega$, or[13.8]

$$\begin{aligned} x(t) &= x_0\, e^{K} \\ &\approx x_0 \exp\,(\pm\, i\,\omega t - \gamma t) \\ &= x_0\, e^{\pm i\,\omega t}.\, e^{-\gamma t} \end{aligned}$$

where $\gamma \equiv \frac{1}{2}k$. We may then think of $[i\,K]$ as a *complex frequency* whose real part is $+\omega$ and whose imaginary part is γ. What sort of situation does this describe? It describes a *weakly damped harmonic motion* in which the usual sinusoidal pattern damps away within an "envelope" whose shape is that of an exponential decay.

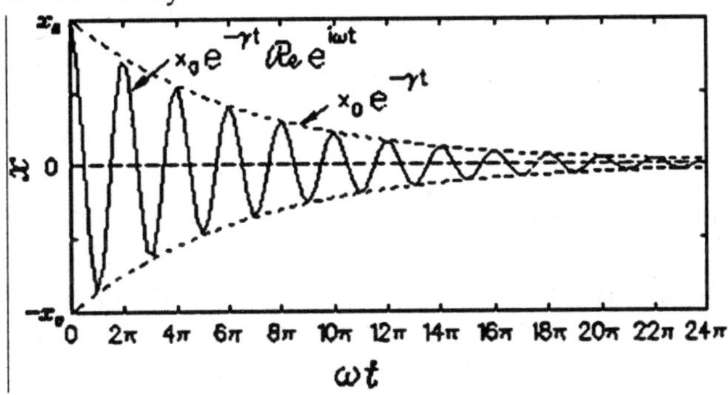

Fig. Weakly Damped Harmonic Motion.

The Initial Amplitude of x (Whatever x is) is x_0, the Angular Frequency is ω and the Damping rate is γ. The Cosine-Like Oscillation, Equivalent to the real part of $x_0\, e^{i\omega t}$, Decays within the Envelope Function $x_0 e^{-\gamma t}$.

GENERALIZATION OF SHM

As for all the other types of *equations of motion, SHM* need not have anything to do with masses, springs or even Physics. Even within Physics, however, there are so many different kinds of examples of *SHM* that we go out of our way to generalize the results: using "q" to represent the "coordinate" whose displacement from the equilibrium "position" (always taken as $q = 0$) engenders some sort of restoring "force" $Q = -kq$ and "μ" to represent an "inertial factor" that plays the rôle of the mass, we have

$$\ddot{q} = -\left(\frac{k}{\mu}\right)q$$

for which the solution is the real part of

$$q(t) = q_0 e^{i\omega t} \text{ where } \sqrt{\frac{k}{\mu}}$$

When some form of "drag" acts on the system, we expect to see the qualitative behaviour pictured. Although one might expect virtually every real example to have some sort of frictional damping term, in fact there are numerous physical examples with no damping whatsoever, mostly from the microscopic world of solids.

THE UNIVERSALITY OF SHM

Why is *SHM* characteristic of such an enormous variety of phenomena? Because *for sufficiently small* displacements from equilibrium, *every* system with an equilibrium configuration satisfies the first condition for *SHM*: the linear restoring force. Here is the simple argument: a linear restoring force is equivalent to a potential energy of the form $V(q) = \frac{1}{2} k\, q^2$ — *i.e.* a "quadratic minimum" of the potential energy at the equilibrium configuration $q = 0$. But if we "blow up" a graph of $V(q)$ near $q = 0$, *every* minimum looks quadratic under sufficient magnification!

That means *any* system that *has* an equilibrium configuration also has some analogue of a "potential energy" which is a minimum there; if it also has some form of *inertia* so that it tends to stay at rest (or in motion) until acted upon by the analogue of a *force*, then it will automatically exhibit *SHM* for small-amplitude displacements. This makes *SHM* an extremely powerful paradigm.

Forced Vibration

Musical instruments and other objects are set into vibration at their natural frequency when a person hits, strikes, strums, plucks or somehow disturbs the object. For instance, a guitar string is strummed or plucked; a piano string is hit with a hammer when a pedal is played; and the tines of a tuning fork are hit with a rubber mallet. Whatever the case, a person or thing puts energy into the instrument by direct contact with it. This input of energy disturbs the particles and forces the object into vibrational motion – at its natural frequency.

If you were to take a guitar string and stretch it to a given length and a given tightness and have a friend pluck it, you would hear a noise; but the noise would not even be close in comparison to the loudness produced by an acoustic guitar. On the other hand, if the string is attached to the sound box of the guitar, the vibrating string is capable of forcing the sound box into vibrating at that same natural frequency. The sound box in turn forces air particles inside the box into vibrational motion at the same natural frequency as the string.

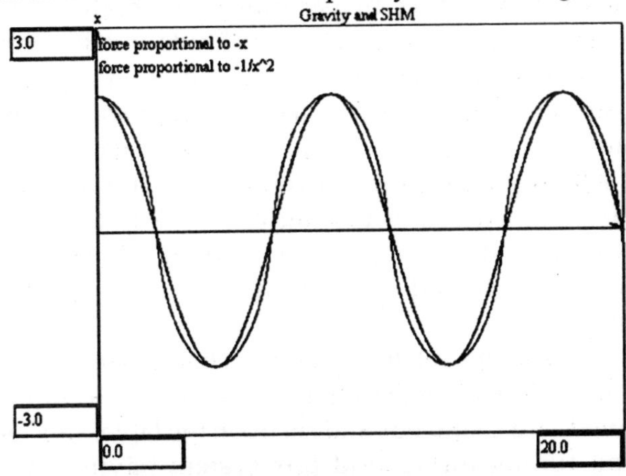

Fig. Force Vibration in SHM

The entire system (string, guitar, and enclosed air) begins vibrating and forces surrounding air particles into vibrational motion. The tendency of one object to force another *adjoining* or *interconnected* object into vibrational motion is referred to as a forced vibration. In the case of the guitar string mounted to the sound box, the fact that the surface area of the sound box is greater than the surface area of the string, means that more surrounding air particles will be forced into vibration. This causes an increase in the amplitude and thus loudness of the sound.

This same principle of a forced vibration is often demonstrated in a Physics classroom using a tuning fork. If the tuning fork is held in your hand and hit with a rubber mallet, a sound is produced as the tines of the tuning fork set surrounding air particles into vibrational motion. The sound produced by the tuning fork is barely audible to students in the back rows of the room.

However, if the tuning fork is set upon the whiteboard panel or the glass panel of the overhead projector, the panel begins vibrating at the same natural frequency of the tuning fork. The tuning fork forces surrounding glass (or vinyl) particles into vibrational motion.

The vibrating whiteboard or overhead projector panel in turn forces surrounding air particles into vibrational motion and the result is an increase in the amplitude and thus loudness of the sound. This principle of forced vibration explains why demonstration tuning forks are mounted on a sound box, why a commercial music box mechanism is mounted on a sounding board, why a guitar utilizes a sound box, and why a piano string is attached to a sounding board. A louder sound is always produced when an accompanying object of greater surface area is forced into vibration at the same natural frequency.

Now consider a related situation which resembles another common Physics demonstration. Suppose that a tuning fork is mounted on a sound box and set upon the table; and suppose a second tuning fork/sound box system having the same natural frequency (say 256 Hz) is placed on the table near the first system. Neither of the tuning forks is vibrating. Suppose the first tuning fork is struck with a rubber mallet and the tines begin vibrating at its natural frequency - 256 Hz. These vibrations set its sound box and the air inside the sound box vibrating at the same natural frequency of 256 Hz. Surrounding air particles are set into vibrational motion at the same natural frequency of 256 Hz and every student in the classroom hears the sound. Then the tines of the tuning fork are grabbed to prevent their vibration and remarkably the sound of 256 Hz is still being heard. Only now the sound is being produced by the second tuning fork - the one which wasn't hit with the mallet. The demonstration is often repeated to assure that the same surprising results are observed. They are! What is happening?

In this demonstration, one tuning fork forces another tuning fork into vibrational motion at the same natural frequency. The two forks are *connected* by the surrounding air particles. As the air particles surrounding the first fork (and its connected sound box) begin vibrating, the pressure waves which it creates begin to impinge at a periodic and regular rate of 256 Hz upon the second tuning fork (and its connected sound box).

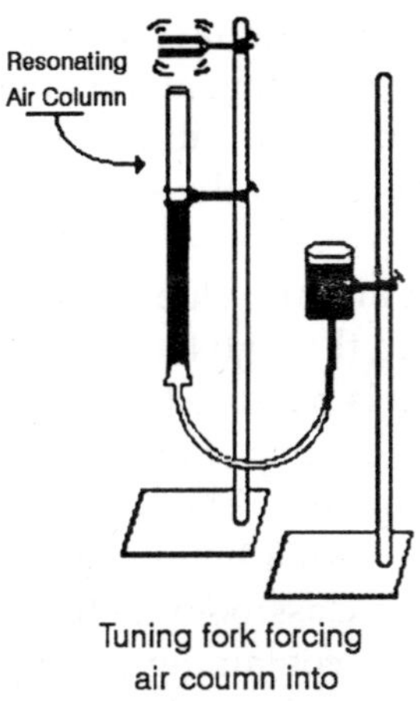

Tuning fork forcing air coumn into resonance

The energy carried by this sound wave through the air is *tuned* to the frequency of the second tuning fork. Since the incoming sound waves share the same natural frequency as the second tuning fork, the tuning fork easily begins vibrating at its natural frequency. This is an example of resonance - when one object vibrating at the same natural frequency of a second object forces that second object into vibrational motion.

The result of resonance is always a large vibration. Regardless of the vibrating system, if resonance occurs, a large vibration results. This is often demonstrated in a Physics class with an odd-looking mechanical system resembling an inverted pendulum.

The apparatus consists of three sets of two identical plastic bobs mounted on a very elastic metal pole, which arere in turn mounted to a metal bar. Each metal pole and attached bob has a different length, thus giving it a different natural frequency of vibration. The bobs are often color coded to distinguish between them; they are colored red, blue and green.

The red bobs are mounted on the longer poles and they have the lowest natural frequency of vibration. The blue bobs are mounted on the shorter poles and have the highest natural frequency of vibration.

When the red bob is disturbed, it begins vibrating at its natural frequency. This in turn forces the attached bar to vibrate at the same frequency; and this forces the other attached red bob into vibrating at the same natural frequency. This is resonance - one bob vibrating at a given frequency forcing a second object with the same natural frequency into vibrational motion.

While the green and the blue bobs were disturbed by the vibrations transmitted through the metal bar, only the red bob would resonate. This is because the frequency of the first red bob is tuned to the frequency of the second red bob; they share the same natural frequency. The result is that the second red bob begins vibrating with a huge amplitude.

Another common classroom demonstration of resonance involves a plastic tube containing an air column. The length of the air column was adjusted by raising and lowering a reservoir of water (dyed red).

The raising and lowering of the reservoir adjusts the height of water in the open-air tube, and thus adjusts the length of the air column inside the tube. As the length of the air column is decreased, the natural frequency of the air column is increased. While adjusting the height of the liquid in the tube, a vibrating tuning fork is held above the air column of the tube.

When the natural frequency of the air column is *tuned* to the frequency of the vibrating tuning fork, resonance occurs and a loud sound results. Quite amazingly, the vibrating tuning fork forces air particles within the air column into vibrational motion. Once more in this resonance situation, the tuning fork and the air column share the same vibrational frequency.

In conclusion, resonance occurs when two interconnected objects share the same vibrational frequency. When one of the objects is vibrating, it forces the second object into vibrational motion. The result is a large vibration. And if a sound wave within the audible range of human hearing is produced, a loud sound is heard.

VIBRATION ANALYSIS

The fundamentals of vibration analysis can be understood by studying the simple mass-spring-damper model. Indeed, even a complex structure such as an automobile body can be modeled as a "summation" of simple mass-spring-damper models.

The mass-spring-damper model is an example of a simple harmonic oscillator. The mathematics used to describe its behavior is identical to other simple harmonic oscillators such as the RLC circuit.

Note: In this article the step by step mathematical derivations will not be included, but will focus on the major equations and concepts in vibration analysis. Please refer to the references at the end of the article for detailed derivations.

Free Vibration Without Damping

To start the investigation of the mass-spring-damper we will assume the damping is negligible and that there is no external force applied to the mass (i.e. free vibration).The force applied to the mass by the spring is proportional to the amount the spring is stretched "x" (assume the spring is already compressed due to the weight of the mass). The proportionality constant, k, is the stiffness of the spring and has units of force/distance (e.g. lbf/in or N/m)

$$F_s = -kx.$$

The force generated by the mass is proportional to the acceleration of the mass as given by Newton's second law of motion.

$$\Sigma F = ma = m\ddot{x} = m\frac{d^2x}{dt^2}.$$

The sum of the forces on the mass then generates this ordinary differential equation:

$$m\ddot{x} = kx = 0.$$

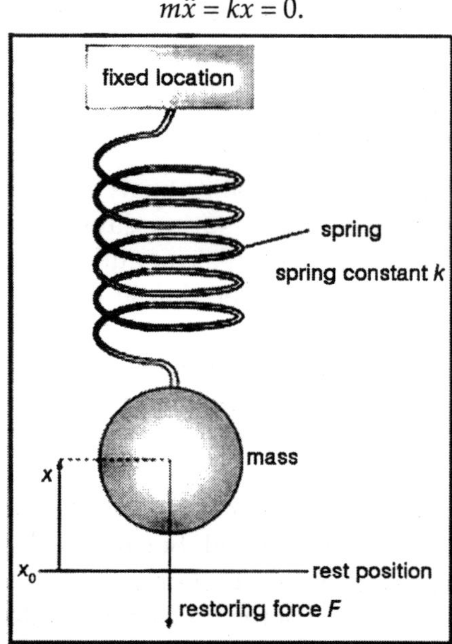

Fig. Simple Harmonic Motion of the Mass-spring System

If we assume that we start the system to vibrate by stretching the spring by the distance of *A* and letting go, the solution to the above equation that describes the motion of mass is:

$$x(t) = A\cos(2\pi f_{nt}).$$

This solution says that it will oscillate with simple harmonic motion that has an amplitude of A and a frequency of f_n. The number f_n is one of the most important quantities in vibration analysis and is called the undamped natural frequency. For the simple mass-spring system, f_n is defined as:

$$f_n = \frac{1}{2\pi}\sqrt{\frac{k}{m}}.$$

Angular frequency ω ($\omega = 2\pi f$) with the units of radians per second is often used in equations because it simplifies the equations, but is normally converted to "standard" frequency (units of Hz or equivalently cycles per second) when stating the frequency of a system.

If you know the mass and stiffness of the system you can determine the frequency at which the system will vibrate once it is set in motion by an initial disturbance using the above stated formula. Every vibrating system has one or more natural frequencies that it will vibrate at once it is disturbed. This simple relation can be used to understand in general what will happen to a more complex system once we add mass or stiffness.

For example, the above formula explains why when a car or truck is fully loaded the suspension will feel "softer" than unloaded because the mass has increased and therefore reduced the natural frequency of the system.

The System to Vibrate Under no Force

These formulas describe the resulting motion, but they do not explain why the system oscillates. The reason for the oscillation is due to the conservation of energy. In the above example we have extended the spring by a value of A and therefore have stored potential energy $\left(\frac{1}{2}kx^2\right)$ in the spring. Once we let go of the spring, the spring tries to return to its un-stretched state and in the process accelerates the mass. At the point where the spring has reached its un-stretched state it no longer has energy stored, but the mass has reached its maximum speed and hence all the energy has been transformed into kinetic energy $\left(\frac{1}{2}mv^2\right)$. The mass then begins to decelerate because it is now compressing the spring and in the process transferring the kinetic energy back to its potential. This transferring back and forth of the kinetic energy in the mass and the potential energy in the spring causes the mass to oscillate.

In our simple model the mass will continue to oscillate forever at the same magnitude, but in a real system there is always something called damping that dissipates the energy and therefore the system eventually bringing it to rest.

Free Vibration with Damping

We now add a "viscous" damper to the model that outputs a force that is proportional to the velocity of the mass. The damping is called viscous because it models the effects of an object within a fluid. The proportionality constant c is called the damping coefficient and has units of Force over velocity (lbf s/ in or N s/m).

$$F_d = -cv = -c\dot{x} = -c\frac{dx}{dt}.$$

By summing the forces on the mass we get the following ordinary differential equation:

$$m\ddot{x} + c\ddot{x} + kx = 0.$$

The solution to this equation depends on the amount of damping. If the damping is small enough the system will still vibrate, but eventually, over time, will stop vibrating. This case is called underdamping—this case is of most interest in vibration analysis. If we increase the damping just to the point where the system no longer oscillates we reach the point of critical damping (if the damping is increased past critical damping the system is called overdamped). The value that the damping coefficient needs to reach for critical damping in the mass spring damper model is:

$$c_c = 2\sqrt{km}.$$

To characterize the amount of damping in a system a ratio called the damping ratio (also known as damping factor and %

critical damping) is used. This damping ratio is just a ratio of the actual damping over the amount of damping required to reach critical damping. The formula for the damping ratio (æ) of the mass spring damper model is:

$$\zeta = \frac{c}{2\sqrt{km}}.$$

For example, metal structures (e.g. airplane fuselage, engine crankshaft) will have damping factors less than 0.05 while automotive suspensions in the range of 0.2-0.3.

The solution to the underdamped system for the mass spring damper model is the following:

$$x(t) = Xe^{-z\omega_n t}\cos\left(\sqrt{1-\zeta^2}\,\omega_n t - \phi\right), \omega_n = 2\pi f_n$$

$Xe^{-\zeta\omega_n t}$

$f_n = 1$

$\omega_n = 2\pi f_n$

$\zeta = 0.1$

Time

$Xe^{-\zeta\omega_n t}$

$f_n = 1$

$\omega_n = 2\pi f_n$

$\zeta = 0.3$

Time

The value of X, the initial magnitude, and ö, the phase shift, are determined by the amount the spring is stretched. The formulas for these values can be found in the references.The major points to note from the solution are the exponential term and the cosine function. The exponential term defines how quickly the system "damps" down – the larger the damping ratio, the quicker it damps to zero. The cosine function is the oscillating portion of the solution, but the frequency of the oscillations is different from the undamped case.The frequency in this case is called the "damped natural frequency", f_d, and is related to the undamped natural frequency by the following formula:

$$f_d = \sqrt{1-\zeta^2} f_n.$$

The damped natural frequency is less than the undamped natural frequency, but for many practical cases the damping ratio is relatively small and hence the difference is negligible. Therefore the damped and undamped description are often dropped when stating the natural frequency (e.g. with 0.1 damping ratio, the damped natural frequency is only 1% less than the undamped).The plots to the side present how 0.1 and 0.3 damping ratios effect how the system will "ring" down over time. What is often done in practice is to experimentally measure the free vibration after an impact (for example by a hammer) and then determine the natural frequency of the system by measuring the rate of oscillation as well as the damping ratio by measuring the rate of decay. The natural frequency and damping ratio are not only important in free vibration, but also characterize how a system will behave under forced vibration.

Forced Vibration with Damping

The behavior of the spring mass damper model when we add a harmonic force in the form below. A force of this type could, for example, be generated by a rotating imbalance.

$$F = F_0 \cos(2\pi f t).$$

If we again sum the forces on the mass we get the following ordinary differential equation:

$$m\ddot{x} + c\dot{x} + kx = F_0 \cos(2\pi ft).$$

The steady state solution of this problem can be written as:

$$x(t) + X\cos(2\pi ft - \phi).$$

The result states that the mass will oscillate at the same frequency, f, of the applied force, but with a phase shift).

The amplitude of the vibration "X" is defined by the following formula.

$$X = \frac{F_0}{k} \frac{1}{\sqrt{(1-r^2)^2 + (2\zeta r)^2}}.$$

Where "r" is defined as the ratio of the harmonic force frequency over the undamped natural frequency of the mass-spring-damper model.

$$r = \frac{f}{f_n}.$$

The phase shift , ö, is defined by following formula.

$$\phi = \arctan\left(\frac{2\zeta r}{1-r^2}\right).$$

The plot of these functions, called "the frequency response of the system", presents one of the most important features in forced vibration. In a lightly damped system when the forcing frequency nears the natural frequency ($r \approx 1$) the amplitude of the vibration can get extremely high. This phenomenon is called resonance (subsequently the natural frequency of a system is often referred to as the resonant frequency). In rotor bearing systems any rotational speed that excites a resonant frequency is referred to as a critical speed.

If resonance occurs in a mechanical system it can be very harmful—leading to eventual failure of the system. Consequently, one of the major reasons for vibration analysis is to predict when this type of resonance may occur and then to determine what steps to take to prevent it from occurring. As the amplitude plot shows, adding damping can significantly reduce the magnitude of the vibration. Also, the magnitude can be reduced if the natural frequency can be shifted away from the forcing frequency by changing the stiffness or mass of the system. If the system cannot be changed, perhaps the forcing frequency can be shifted (for example, changing the speed of the machine generating the force).

The following are some other points in regards to the forced vibration shown in the frequency response plots.

- At a given frequency ratio, the amplitude of the vibration, X, is directly proportional to the amplitude of the force F_0 (e.g. If you double the force, the vibration doubles)
- With little or no damping, the vibration is in phase with the forcing frequency when the frequency ratio $r < 1$ and 180 degrees out of phase when the frequency ratio $r > 1$
- When r<<1 the amplitude is just the deflection of the spring under the static force F_0. This deflection is called the static deflection δ_{st}. Hence, when r<<1 the effects of the damper and the mass are minimal.
- When r>>1 the amplitude of the vibration is actually less than the static deflection δ_{st}. In this region the force generated by the mass (F=ma) is dominating because the acceleration seen by the mass increases with the frequency. Since the deflection seen in the spring, X, is reduced in this region, the force transmitted by the spring ($F=kx$) to the base is reduced. Therefore the mass-spring-damper system is isolating the harmonic force from the mounting base—referred to as vibration isolation. Interestingly, more damping actually reduces the effects of vibration isolation when r>>1 because the damping force ($F=cv$) is also transmitted to the base.

Causes of Resonance

Resonance is simple to understand if you view the spring and mass as energy storage elements—with the mass storing kinetic energy and the spring storing potential energy. When the mass and spring have no force acting on them they transfer energy back forth at a rate equal to the natural frequency. In other words, if energy is to be efficiently pumped into both the mass and spring the energy source needs to feed the energy in at a rate equal to the natural frequency. Applying a force to the mass and spring is similar to pushing a child on swing, you need to push at the correct moment if you want the swing to get higher and higher. As in the case of the swing, the force applied does not necessarily have to be high to get large motions; the pushes just need to keep adding energy into the system.

The damper, instead of storing energy, dissipates energy. Since the damping force is proportional to the velocity, the more the motion the more the damper dissipates the energy. Therefore a point will come when the energy dissipated by the damper will equal the energy being fed in by the force. At this point, the system has reached its maximum amplitude and will continue to vibrate at this level as long as the force applied stays the same. If no damping exists, there is nothing to dissipate the energy and therefore theoretically the motion will continue to grow on into infinity.

COMPLEX FORCES TO THE MASS-SPRING-DAMPER MODEL

A simple harmonic force was applied to the model, but this can be extended considerably using two powerful mathematical tools. The first is the Fourier transform that takes a signal as a function of time (time domain) and breaks it down into its harmonic components as a function of frequency (frequency domain). For example, let us apply a force to the mass-spring-damper model that repeats the following cycle—a force equal to 1 newton for 0.5 second and then no force for 0.5 second. This type of force has the shape of a 1 Hz square wave.

The Fourier transform of the square wave generates a frequency spectrum that presents the magnitude of the harmonics that make up the square wave (the phase is also generated, but is typically of less concern and therefore is often not plotted). The Fourier transform can also be used to analyze non-periodic functions such as transients (e.g. impulses) and random functions. With the advent of the modern computer the Fourier transform is almost always computed using the Fast Fourier Transform (FFT) computer algorithm in combination with a window function.

In the case of our square wave force, the first component is actually a constant force of 0.5 newton and is represented by a value at "0" Hz in the frequency spectrum. The next component is a 1 Hz sine wave with an amplitude of 0.64. This is shown by the line at 1 Hz. The remaining components are at odd frequencies and it takes an infinite amount of sine waves to generate the perfect square wave. Hence, the Fourier transform allows you to interpret the force as a sum of sinusoidal forces being applied instead of a more "complex" force (e.g. a square wave).

The vibration solution was given for a single harmonic force, but the Fourier transform will in general give multiple harmonic forces. The second mathematical tool, "the principle of superposition", allows you to sum the solutions from multiple forces if the system is linear. In the case of the spring-mass-damper model, the system is linear if the spring force is proportional to the displacement and the damping is proportional to the velocity over the range of motion of interest. Hence, the solution to the problem with a square wave is summing the predicted vibration from each one of the harmonic forces found in the frequency spectrum of the square wave.

Frequency Response Model

We can view the solution of a vibration problem as an input/output relation—where the force is the input and the output is the vibration. If we represent the force and vibration in the frequency domain (magnitude and phase) we can write the following relation:

$$X(\omega) = H(\omega) * F(\omega) \text{ or } H(\omega) = \frac{X(\omega)}{F(\omega)}.$$

$H(\omega)$ is called the frequency response function (also referred to the transfer function, but not technically as accurate) and has both a magnitude and phase component (if represented as a complex number, a real and imaginary component). The magnitude of the frequency response function (FRF) was presented earlier for the mass-spring-damper system.

$$|H(\omega)| = \left|\frac{X(\omega)}{F(\omega)}\right| = \frac{1}{k}\frac{1}{\sqrt{(1-r^2)^2 + (2\zeta r)^2}}, \text{ where } r = \frac{f}{f_n} = \frac{\omega}{\omega_n}.$$

The phase of the FRF was also presented earlier as:

$$\angle H(\omega) = \arctan\left(\frac{2\zeta r}{1-r^2}\right).$$

For example, let us calculate the FRF for a mass-spring-damper system with a mass of 1 kg, spring stiffness of 1.93 N/mm and a damping ratio of 0.1. The values of the spring and mass give a natural frequency of 7 Hz for this specific system. If we apply the 1 Hz square wave from earlier we can calculate the predicted vibration of the mass. The resulting vibration. It happens in this example that the fourth harmonic of the square wave falls at 7 Hz. The frequency response of the mass-spring-damper therefore outputs a high 7 Hz vibration even though the input force had a relatively low 7 Hz harmonic. This example highlights that the resulting vibration is dependent on both the forcing function and the system that the force is applied.

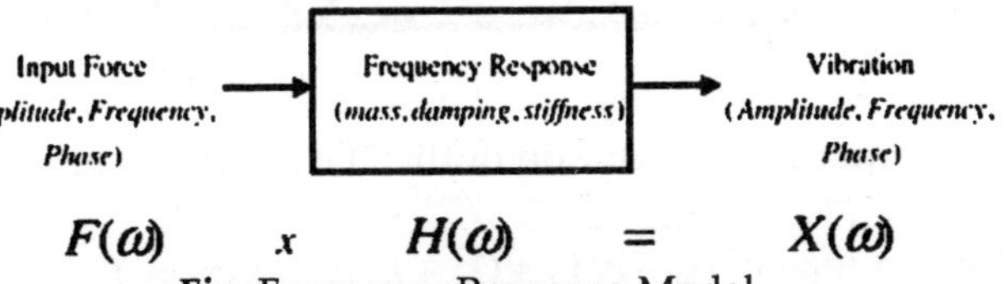

Input Force (*Amplitude, Frequency, Phase*)	→	Frequency Response (*mass, damping, stiffness*)	→	Vibration (*Amplitude, Frequency, Phase*)
$F(\omega)$	x	$H(\omega)$	=	$X(\omega)$

Fig. Frequency Response Model.

The figure also shows the time domain representation of the resulting vibration. This is done by performing an inverse Fourier Transform that converts frequency domain data to time domain. In practice, this is rarely done because the frequency spectrum provides all the necessary information.

The frequency response function (FRF) does not necessarily have to be calculated from the knowledge of the mass, damping, and stiffness of the system, but can be measured experimentally. For example, if you apply a known force and sweep the frequency and then measure the resulting vibration you can calculate the frequency response function and then characterize the system. This technique is used in the field of experimental modal analysis to determine the vibration characteristics of a structure.

MULTIPLE DEGREES OF FREEDOM SYSTEMS AND MODE SHAPES

The simple mass-spring damper model is the foundation of vibration analysis, but what about more complex systems? The mass-spring-damper model described above is called a single degree of freedom (DOF) model since we have assumed the mass only moves up and down. In the case of more complex systems we need to discretize the system into more masses and allow them to move in more than one direction—adding degrees of freedom. The major concepts of multiple degrees of freedom (MDOF) can be understood by looking at just a 2 degree of freedom model.

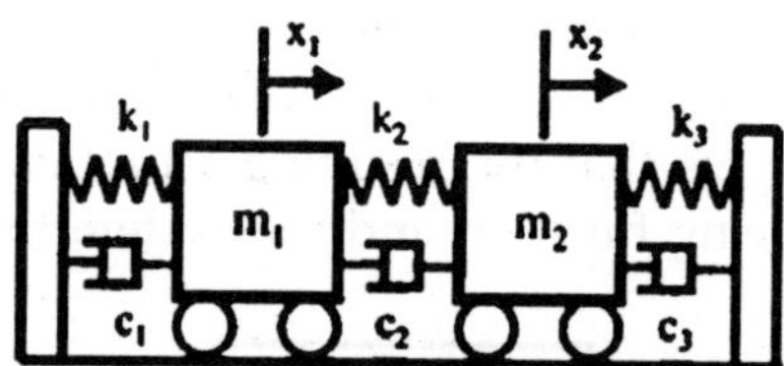

Fig. Degree of Freedom Model

The equations of motion of the 2DOF system are found to be:

$$m_1\ddot{x}_1 + (c_1 + c_2)\dot{x}_1 - c_2\dot{x}_2 + (k_1 + k_2)x_1 - k_2x_2 = f_1$$

$$m_2\ddot{x}_2 + (c_2\dot{x}_1 + (c_2 + c_3)\dot{x}_2 - c_2\dot{x}_3 - k_2x_1 + (k_2 + k_3) - k_2x_3 = f_2.$$

We can rewrite this in matrix format:

$$\begin{bmatrix} m_1 & 0 \\ 0 & m_2 \end{bmatrix} \begin{Bmatrix} \ddot{x}_1 \\ \ddot{x}_2 \end{Bmatrix} + \begin{bmatrix} c_1 + c_2 & -c_2 \\ -c_2 & c_2 + c_3 \end{bmatrix} \begin{Bmatrix} \dot{x}_1 \\ \dot{x}_2 \end{Bmatrix}$$

$$+\begin{bmatrix} k_1+k_2 & -k_2 \\ -k_2 & k_2+k_3 \end{bmatrix}\begin{Bmatrix} x_1 \\ x_2 \end{Bmatrix}=\begin{Bmatrix} f_1 \\ f_2 \end{Bmatrix}$$

A more compact form of this matrix equation can be written as:

$$[M]\{\ddot{x}\}+[C]\{\dot{x}\}+[K]\{x\}=\{f\}$$

where $[M]$, [C], and [K] are symmetric matrices referred respectively as the mass, damping, and stiffness matrices. The matrices are NxN square matrices where N is the number of degrees of freedom of the system.

In the following analysis we will consider the case where there is no damping and no applied forces (i.e. free vibration). The solution of a viscously damped system is somewhat more complicated and is shown in Maia, Silva..

$$[M]\{\ddot{x}\}+[K]\{x\}=0.$$

This differential equation can be solved by assuming the following type of solution:

$$\{x\}=\{X\}e^{i\omega t}.$$

Note: Using the exponential solution of $\{X\}\,e^{i\omega t}$ is a mathematical trick used to solve linear differential equations. If we use Euler's formula and take only the real part of the solution it is the same cosine solution for the 1 DOF system. The exponential solution is only used because it easier to manipulate mathematically.

The equation then becomes:

$$\left[-\omega^2[M]+[K]\right]\{X\}e^{i\omega t}=0.$$

Since $e^{i\grave{u}t}=0$ cannot equal zero the equation reduces to the following.

$$\left[[K]-\omega^2[M]\right]\{X\}=0.$$

Eigenvalue Problem

This is referred to an eigenvalue problem in mathematics and can be put in the standard format by pre-multiplying the equation by$[M]^{-1}$

$$\left[[M]^{-1}[K]-\omega^2[M]^{-1}[M]\right]\{X\}=0$$

and if we let $[M]^{-1}$ $[K]$ = $[A]$ and $\lambda = \omega^2$

$$[[A]-\lambda[I]]\{X\}=0.$$

The solution to the problem results in N eigenvalues $\left(\text{ie.}\,\omega_1^2,\omega_2^2,..\omega_N^2\right)$, where N corresponds to the number of degrees of freedom. The eigenvalues provide the natural frequencies of the system. When these eigenvalues are substituted back into the original set of equations, the values of $\{X\}$ that correspond to each eigenvalue are called the eigenvectors. These eigenvectors represent the mode shapes of the system. The solution of an eigenvalue problem can be quite cumbersome (especially for problems with many degrees of freedom), but fortunately most math analysis programs have eigenvalue routines.The eigenvalues and eigenvectors are often written in the following matrix format and describe the modal model of the system:

$$\left[{}^{\backslash}\omega_{r\backslash}^2\right]=\begin{bmatrix}\omega_1^2 & \cdots & 0\\ \vdots & \ddots & \vdots\\ 0 & \cdots & \omega_N^2\end{bmatrix} \text{ and } [Y] = \left[\{\psi_1\}\{\psi_2\}\cdots\{\psi_N\}\right]$$

A simple example using our 2 DOF model. Let both masses have a mass of 1 kg and the stiffness of all three springs equal 1000 N/m. The mass and stiffness matrix for this problem are then:

$$[M]=\begin{bmatrix}1 & 0\\ 0 & 1\end{bmatrix}_{\text{and}}\ [K]=\begin{bmatrix}2000 & -1000\\ -1000 & 2000\end{bmatrix}$$

Then

$$[A]=\begin{bmatrix}2000 & -1000\\ -1000 & 2000\end{bmatrix}.$$

The eigenvalues for this problem given by an eigenvalue routine will be:

$$\left[{}^{\backslash}\omega_{r\backslash}^2\right]=\begin{bmatrix}1000 & 0\\ 0 & 3000\end{bmatrix}.$$

The natural frequencies in the units of hertz are then (remembering $\omega = 2\pi f$) $f_1 = 31.62Hz$ and $f_2 = 54.77Hz$.

The two mode shapes for the respective natural frequencies are given as:

$$[\Psi] = [\{\psi_1\}\{\psi_2\}] = \left[\begin{Bmatrix} -0.707 \\ -0.707 \end{Bmatrix}_1 \begin{Bmatrix} 0.707 \\ -0.707 \end{Bmatrix}_2 \right].$$

Since the system is a 2 DOF system, there are two modes with their respective natural frequencies and shapes. The mode shape vectors are not the absolute motion, but just describe relative motion of the degrees of freedom. In our case the first mode shape vector is saying that the masses are moving together in phase since they have the same value and sign. In the case of the second mode shape vector, each mass is moving in opposite direction at the same rate.

Multiple DOF Problem

When there are many degrees of freedom, the best method of visualizing the mode shapes is by animating them. An example of animated mode shapes for a cantilevered I-beam. In this case, a finite element model was used to generate the mass and stiffness matrices and solve the eigenvalue problem. Even this relatively simple model has over a 100 degrees of freedom and hence as many natural frequencies and mode shapes. In general only the first few modes are important.

The mode shapes of a cantilevered I-beam

1st lateral bending	1st torsional	1st vertical bending
2nd lateral bending	2nd torsional	2nd vertical bending

Multiple DOF Problem Converted to a Single DOF Problem

The eigenvectors have very important properties called orthogonality properties. These properties can be used to greatly simplify the solution of multi-degree of freedom models. It can be shown that the eigenvectors have the following properties:

$$[\Psi]^T [M][\Psi] = [\backslash m_{r\backslash}]$$

$$[\Psi]^T [K][\Psi] = [\backslash k_{r\backslash}]$$

$[\backslash m_r \backslash]$ and $[\backslash k_r \backslash]$ are diagonal matrices that contain the modal mass and stiffness values for each one of the modes.

(Note: Since the eigenvectors (mode shapes) can be arbitrarily scaled, the orthogonality properties are often used to scale the eigenvectors so the modal mass value for each mode is equal to 1. The modal mass matrix is therefore an identity matrix)

These properties can be used to greatly simplify the solution of multi-degree of freedom models by making the following the coordinate transformation.

$$\{x\} = [\Psi]\{q\}.$$

If we use this coordinate transformation in our original free vibration differential equation we get the following equation.

$$[M][\Psi]\{\ddot{q}\} + [K][\Psi]\{q\} = 0.$$

We can take advantage of the orthogonality properties by premultiplying this equation by $[\Psi]^T$

$$[\Psi]^T [M][\Psi]\{\ddot{q}\} + [\Psi]^T [K][\Psi]\{q\} = 0.$$

The orthogonality properties then simplify this equation to:

$$[\backslash m_r \backslash][\ddot{q}] + [\backslash k_r \backslash]\{q\} = 0.$$

This equation is the foundation of vibration analysis for multiple degree of freedom systems. A similar type of result can be derived for damped systems. The key is that the modal and stiffness matrices are diagonal matrices and therefore we have "decoupled" the equations.

In other words, we have transformed our problem from a large unwieldy multiple degree of freedom problem into many single degree of freedom problems that can be solved using the same methods outlined above.

Instead of solving for x we are instead solving for q, referred to as the modal coordinates or modal participation factors.

It may be clearer to understand if we write $\{x\} = [\Psi]\{q\}$ as:

$$\{x_n\} = q_1\{\psi\}_1 + q_2\{\psi\}_2 + q_3\{\psi\}_3 + \cdots + q_N\{\psi\}_N.$$

Written in this form we can see that the vibration at each of the degrees of freedom is just a linear sum of the mode shapes. Furthermore, how much each mode "participates" in the final vibration is defined by q, its modal participation factor.

REFRACTIVE INDEX

The refractive index (or index of refraction) of a medium is a measure for how much the speed of light (or other waves such as sound waves) is reduced inside the medium. For example, typical glass has a refractive index of 1.5, which means that in glass, light travels at 1 / 1.5 = 0.67 times the speed of light in a vacuum. Two common properties of glass and other transparent materials are directly related to their refractive index. First, light rays change direction when they cross the interface from air to the material, an effect that is used in lenses and glasses. Second, light reflects partially from surfaces that have a refractive index different from that of their surroundings.

Definition: The refractive index n of a medium is defined as the ratio of the phase velocity c of a wave phenomenon such as light or sound in a reference medium to the phase velocity v_p in the medium itself:

$$n = \frac{c}{v_p}.$$

It is most commonly used in the context of light with vacuum as a reference medium, although historically other reference media (e.g. air at a standardized pressure and temperature) have been common. It is usually given the symbol n. In the case of light, it equals

$$n = \sqrt{\epsilon_r \mu_r},$$

where ε_r is the material's relative permittivity, and μ_r is its relative permeability. For most materials, μ_r is very close to 1 at optical frequencies, therefore n is approximately $\sqrt{\epsilon_r}$. Contrary to a widespread misconception, n may be less than 1, for example for x-rays.

This has practical technical applications, such as effective mirrors for x-rays based on total external reflection. The phase velocity is defined as the rate at which the crests of the waveform propagate; that is, the rate at which the phase of the waveform is moving.

The *group velocity* is the rate that the *envelope* of the waveform is propagating; that is, the rate of variation of the amplitude of the waveform. Provided the waveform is not distorted significantly during propagation, it is the group velocity that represents the rate that information (and energy) may be transmitted by the wave, for example the velocity at which a pulse of light travels down an optical fiber.

Speed of Light

Refraction of light at the interface between two media of different refractive indices, with $n_2 > n_1$. Since the phase velocity is lower in the second medium ($v_2 < v_1$), the angle of refraction $è_2$ is less than the angle of incidence $è_1$; that is, the ray in the higher-index medium is closer to the normal.

The speed of all electromagnetic radiation in vacuum is the same, approximately 3×10^8 meters per second, and is denoted by *c*. Therefore, if *v* is the phase velocity of radiation of a specific frequency in a specific material, the refractive index is given by

$$n = \frac{c}{v}$$

or inversely

$$v = \frac{c}{n}.$$

This number is typically greater than one: the higher the index of the material, the more the light is slowed down. However, at certain frequencies (e.g. near absorption resonances, and for X-rays), *n* will actually be smaller than one. This does not contradict the theory of relativity, which holds that no information-carrying signal can ever propagate faster than *c*, because the phase velocity is not the same as the group velocity or the signal velocity.

Sometimes, a "group velocity refractive index", usually called the *group index* is defined:

$$n_g = \frac{c}{v_g}$$

where v_g is the group velocity. This value should not be confused with n, which is always defined with respect to the phase velocity. The group index can be written in terms of the wavelength dependence of the refractive index as

$$n_g = n - \lambda \frac{dn}{d\lambda},$$

where λ is the wavelength in vacuum. At the microscale, an electromagnetic wave's phase velocity is slowed in a material because the electric field creates a disturbance in the charges of each atom (primarily the electrons) proportional to the permittivity of the medium. The charges will, in general, oscillate slightly out of phase with respect to the driving electric field. The charges thus radiate their own electromagnetic wave that is at the same frequency but with a phase delay.

The macroscopic sum of all such contributions in the material is a wave with the same frequency but shorter wavelength than the original, leading to a slowing of the wave's phase velocity. Most of the radiation from oscillating material charges will modify the incoming wave, changing its velocity. However, some net energy will be radiated in other directions. If the refractive indices of two materials are known for a given frequency, then one can compute the angle by which radiation of that frequency will be refracted as it moves from the first into the second material from Snell's law. If in a given region the values of refractive indices n or n_g were found to differ from unity (whether homogeneously, or isotropically, or not), then this region was distinct from vacuum in the above sense for lacking Poincaré symmetry.

Negative Refractive Index

Recent research has also demonstrated the existence of negative refractive index which can occur if the real parts of both ε_r and μ_r are *simultaneously* negative, although such is a necessary but not sufficient condition. Not thought to occur naturally, this can be achieved with so-called metamaterials and offers the possibility of perfect lenses and other exotic phenomena such as a reversal of Snell's law.

DISPERSION AND ABSORPTION

In real materials, the polarization does not respond instantaneously to an applied field. This causes dielectric loss, which can be expressed by a permittivity that is both complex and frequency dependent. Real materials are not perfect insulators either, i.e. they have non-zero direct current conductivity. Taking both aspects into consideration, we can define a complex index of refraction:

$$\tilde{n} = n + ik.$$

Here, n is the refractive index indicating the phase velocity as above, while k is called the extinction coefficient, which indicates the amount of absorption loss when the electromagnetic wave propagates through the material. Both n and k are dependent on the frequency (wavelength). Note that the sign of the complex part is a matter of convention, which is important due to possible confusion between loss and gain. The notation above, which is usually used by physicists, corresponds to waves with time evolution given by $e^{-i\omega t}$.

The effect that n varies with frequency (except in vacuum, where all frequencies travel at the same speed, c) is known as dispersion, and it is what causes a prism to divide white light into its constituent spectral colors, explains rainbows, and is the cause of chromatic aberration in lenses. In regions of the spectrum where the material does not absorb, the real part of the refractive index tends to increase with frequency. Near absorption peaks, the curve of the refractive index is a complex form given by the Kramers–Kronig relations, and can decrease with frequency.

Since the refractive index of a material varies with the frequency (and thus wavelength) of light, it is usual to specify the corresponding vacuum wavelength at which the refractive index is measured. Typically, this is done at various well-defined spectral emission lines; for example, n_D is the refractive index at the Fraunhofer "D" line, the centre of the yellow sodium double emission at 589.29 nm wavelength.The Sellmeier equation is an empirical formula that works well in describing dispersion, and Sellmeier coefficients are often quoted instead of the refractive index in tables. For some representative refractive indices at different wavelengths.

Dielectric loss and non-zero DC conductivity in materials cause absorption. Good dielectric materials such as glass have extremely low DC conductivity, and at low frequencies the dielectric loss is also negligible, resulting in almost no absorption ($k \approx 0$). However, at higher frequencies (such as visible light), dielectric loss may increase absorption significantly, reducing the material's transparency to these frequencies.The real and imaginary parts of the complex refractive index are related through use of the Kramers–Kronig relations. For example, one can determine a material's full complex refractive index as a function of wavelength from an absorption spectrum of the material.

Relation to Dielectric Constant

The dielectric constant (which is often dependent on wavelength) is simply the square of the (complex) refractive index. The refractive index is used for optics in Fresnel equations and Snell's law; while the dielectric constant is used in Maxwell's equations and electronics.

Where $\tilde{\epsilon}$, ε_1, ε_2, n, and κ are functions of wavelength:

$$\tilde{\epsilon} = \epsilon_1 + i\,\epsilon_2 = (n + i\kappa)^2.$$

Conversion between refractive index and dielectric constant is done by:

$$\varepsilon_1 = n^2 - \kappa^2$$

$$\varepsilon_2 = 2n\kappa$$

$$n = \sqrt{\frac{\sqrt{\epsilon_1^2 + \epsilon_2^2} + \epsilon_1}{2}}$$

$$\kappa = \sqrt{\frac{\sqrt{\epsilon_1^2 + \epsilon_2^2} + \epsilon_1}{2}}$$

Anisotropy

The refractive index of certain media may be different depending on the polarization and direction of propagation of the light through the medium. This is known as birefringence or anisotropy and is described by the field of crystal optics.

In the most general case, the *dielectric constant* is a rank–2 tensor (a 3 by 3 matrix), which cannot simply be described by refractive indices except for polarizations along principal axes.

In magneto-optic (gyro-magnetic) and optically active materials, the principal axes are complex (corresponding to elliptical polarizations), and the dielectric tensor is complex-Hermitian (for lossless media); such materials break time-reversal symmetry and are used e.g. to construct Faraday isolators.

Nonlinearity

The strong electric field of high intensity light (such as output of a laser) may cause a medium's refractive index to vary as the light passes through it, giving rise to nonlinear optics.

If the index varies quadratically with the field (linearly with the intensity), it is called the optical Kerr effect and causes phenomena such as self-focusing and self-phase modulation. If the index varies linearly with the field (which is only possible in materials that do not possess inversion symmetry), it is known as the Pockels effect.

Inhomogeneity

If the refractive index of a medium is not constant, but varies gradually with position, the material is known as a gradient-index medium and is described by gradient index optics.

Light travelling through such a medium can be bent or focussed, and this effect can be exploited to produce lenses, some optical fibers and other devices. Some common mirages are caused by a spatially-varying refractive index of air.

Relation to Density

In general, the refractive index of a glass increases with its density. However, there does not exist an overall linear relation between the refractive index and the density for all silicate and borosilicate glasses. A relatively high refractive index and low density can be obtained with glasses containing light metal oxides such as Li_2O and MgO, while the opposite trend is observed with glasses containing PbO and BaO.

Momentum Paradox

The momentum of a refracted ray, *p*, was calculated by Hermann Minkowski in 1908, where *E* is energy of the photon, *c* is the speed of light in vacuum and *n* is the refractive index of the medium.

$$p = \frac{nE}{c}.$$

RAYLEIGH SCATTERING

Let us now consider the scattering of electromagnetic radiation by neutral atoms. For instance, consider a hydrogen atom. The atom consists of a light electron and a massive proton. The electron scatters radiation much more strongly than the proton, so let us concentrate on the response of the electron to an incident electromagnetic wave. Suppose that the wave electric field is again polarized in the -direction, and is given by Eq We can approximate the electron's equation of motion as

$$m_e \frac{d^2 z}{dt^2} = -m_e \omega_0^2 z - e\, E_0 \sin(\omega t).$$

Here, the second term on the right-hand side represents the perturbing force due to the electromagnetic wave, whereas the first term represents the (linearized) force of electrostatic attraction between the electron and the proton. Indeed, we are very crudely modeling our hydrogen atom as a *simple harmonic oscillator* of natural frequency ω_0. We can think of ω_0 as the typical frequency of electromagnetic radiation emitted by the atom after it is transiently disturbed. In other words, in our model, ω_0 should match the frequency of one of the spectral lines of hydrogen. More generally, we can extend the above model to deal with just about any type of atom, provided that we set ω_0 to the frequency of a spectral line.

We can easily solve Eq. to give

$$z = \frac{e\,E_0}{m_e\left(\omega^2 - \omega_0^2\right)} \sin(\omega t).$$

Hence, the dipole moment of the electron takes the form $\mathbf{p} = -\,p_0 \sin(\omega t)\,\hat{z}$, where

$$p_0 = \frac{e^2\,E_0}{m_e\left(\omega^2 - \omega_0^2\right)}.$$

It follows, by analogy with the analysis in the that the differential and total scattering cross-sections of our model atom take the form

$$\frac{d\sigma}{d\Omega} = \frac{\omega^4}{\left(\omega^2 - \omega_0^2\right)^2} \tau_e^2 \sin^2\theta,$$

and

$$\sigma = \frac{\omega^4}{\left(\omega^2 - \omega_0^2\right)^2} \sigma_T,$$

respectively.

In the limit in which the frequency of the incident radiation is much *greater* than the natural frequency of the atom, Eqs.reduce to the previously obtained expressions for scattering by a free electron. In other words, an electron in an atom acts very much like a free electron as far as high frequency radiation is concerned. In the opposite limit, in which the frequency of the incident radiation is much *less* than the natural frequency of the atom, Eqs. yield

$$\frac{d\sigma}{d\Omega} = \left(\frac{\omega}{\omega_0}\right)^4 \tau_e^2 \sin^2\theta,$$

and

$$\sigma = \left(\frac{\omega}{\omega_0} \right)^4 \sigma_T,$$

respectively. This type of scattering is called *Rayleigh scattering*. There are two features of Rayleigh scattering which are worth noting. First of all, it is much weaker than Thompson scattering (since $\omega << \omega_0$). Secondly, unlike Thompson scattering, it is highly *frequency dependent*. Indeed, it is clear, from the above formulae, that high frequency (short wave-length) radiation is scattered far more effectively than low frequency (long wave-length) radiation.

The most common example of Rayleigh scattering is the scattering of visible radiation from the Sun by neutral atoms (mostly Nitrogen and Oxygen) in the upper atmosphere. The frequency of visible radiation is much less than the typical emission frequencies of a Nitrogen or Oxygen atom (which lie in the ultra-violet band), so it is certainly the case that $\omega << \omega_0$. When the Sun is low in the sky, radiation from it has to traverse a comparatively long path through the atmosphere before reaching us. Under these circumstances, the scattering of direct solar light by neutral atoms in the atmosphere becomes noticeable (it is not noticeable when the Sun is high is the sky, and radiation from it consequently only has to traverse a relatively short path through the atmosphere before reaching us).

According to Eq.blue light is scattered slightly more strongly than red light (since blue light has a slightly higher frequency than red light). Hence, when the Sun is low in the sky, it appears less bright, due to atmospheric scattering. However, it also appears *redder* than normal, because more blue light than red light is scattered out of the solar light-rays, leaving an excess of red light.

Likewise, when we look up at the sky, it does not appear black (like the sky on the Moon) because of light from solar radiation which grazes the atmosphere being scattered downward towards the surface of the Earth. Again, since blue light is scattered more effectively than red light, there is an excess of blue light scattered downward, and so the sky appears *blue.*

Light from the Sun is unpolarized. However, when it is scattered it becomes polarized, because light is scattered preferentially in some directions rather than others. Consider a light-ray from the Sun which grazes the Earth's atmosphere. The light-ray contains light which is polarized such that the electric field is *vertical* to the ground, and light which is polarized such that the electric field is *horizontal* to the ground (and perpendicular to the path of the light-ray), in equal amounts.

However, due to the $\sin^2 \theta$ factor in the dipole emission formula (where, in this case, θ is the angle between the direction of the wave electric field and the direction of scattering), very little light is scattered downward from the vertically polarized light compared to the horizontally polarized light.

Moreover, the light scattered from the horizontally polarization is such that its electric field is preferentially perpendicular, rather than parallel, to the direction of propagation of the solar light-ray (*i.e.,* the direction to the Sun). Consequently, the blue light from the sky is preferentially polarized in a direction *perpendicular* to the direction to the Sun.

Chapter 5

Paraxial Optics

SPHERICAL MIRRORS

A spherical mirror is a mirror which has the shape of a piece cut out of a spherical surface. There are two types of spherical mirrors: *concave,* and *convex*. The most commonly occurring examples of concave mirrors are shaving mirrors and makeup mirrors. As is well-known, these types of mirrors magnify objects placed close to them. The most commonly occurring examples of convex mirrors are the passenger-side wing mirrors of cars. These type of mirrors have wider fields of view than equivalent flat mirrors, but objects which appear in them generally look smaller (and, therefore, farther away) than they actually are.

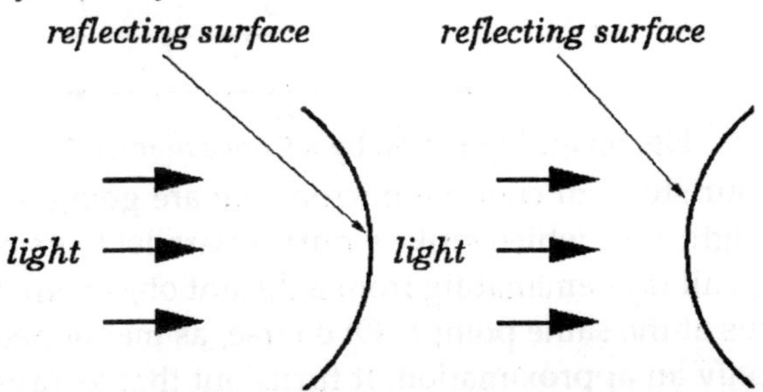

Fig. A Concave (Left) and a Convex (Right) Mirror

Let us now introduce a few key concepts which are needed to study image formation by a concave spherical mirror. The normal to the centre of the mirror is called the *principal axis*. The mirror is assumed to be *rotationally symmetric* about this axis. Hence, we can represent a three-dimensional mirror in a two-dimensional diagram, without loss of generality. The point *V* at which the principal axis touches the surface of the mirror is called the *vertex*.

The point C, on the principal axis, which is equidistant from all points on the reflecting surface of the mirror is called the *centre of curvature*. The distance along the principal axis from point Cto point *V* is called the *radius of curvature* of the mirror, and is denoted *R*. It is found experimentally that rays striking a concave mirror parallel to its principal axis, and not too far away from this axis, are reflected by the mirror such that they all pass through the same point *F* on the principal axis.

This point, which is lies between the centre of curvature and the vertex, is called the *focal point*, or *focus*, of the mirror. The distance along the principal axis from the focus to the vertex is called the *focal length* of the mirror, and is denoted *f*.

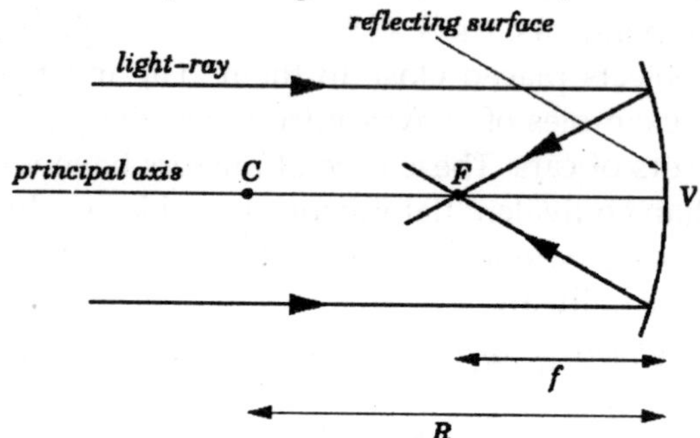

Fig. Image Formation by a Concave mirror.

In our study of concave mirrors, we are going to assume that all light-rays which strike a mirror parallel to its principal axis (*e.g.*, all rays emanating from a distant object) are brought to a focus at the same point *F*. Of course, as mentioned above, this is only an approximation. It turns out that as rays from a distant object depart further from the principal axis of a concave mirror they are brought to a focus ever closer to the mirror.

This lack of perfect focusing of a spherical mirror is called *spherical aberration*. The approximation in which we neglect spherical aberration is called the *paraxial approximation*.[3]

Likewise, the study of image formation under this approximation is known as *paraxial optics*. This field of optics was first investigated systematically by the famous German mathematician Karl Friedrich Gauss in 1841.

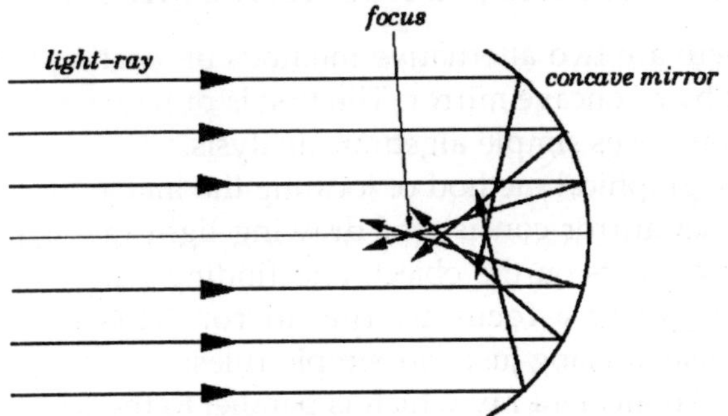

Fig. Spherical Aberration in a Concave Mirror.

It can be demonstrated, by geometry, that the only type of mirror which does not suffer from spherical aberration is a *parabolic* mirror (*i.e.*, a mirror whose reflecting surface is the surface of revolution of a parabola).

Thus, a ray traveling parallel to the principal axis of a parabolic mirror is brought to a focus at the same point *F*, no matter how far the ray is from the axis. Since the path of a light-ray is completely *reversible*, it follows that a light source placed at the focus *F*of a parabolic mirror yields a perfectly parallel beam of light, after the light has reflected off the surface of the mirror. Parabolic mirrors are more difficult, and, therefore, more expensive, to make than spherical mirrors. Thus, parabolic mirrors are only used in situations where the spherical aberration of a conventional spherical mirror would be a serious problem.

The receiving dishes of radio telescopes are generally parabolic. They reflect the incoming radio waves from (very) distant astronomical sources, and bring them to a focus at a single point, where a detector is placed. In this case, since the sources are extremely faint, it is imperative to avoid the signal losses which would be associated with spherical aberration. A car headlight consists of a light-bulb placed at the focus of a parabolic reflector. The use of a parabolic reflector enables the headlight to cast a very straight beam of light ahead of the car. The beam would be nowhere near as well-focused were a spherical reflector used instead.

IMAGE FORMATION BY CONCAVE MIRRORS

There are two alternative methods of locating the image formed by a concave mirror. The first is purely graphical, and the second uses simple algebraic analysis.

The graphical method of locating the image produced by a concave mirror consists of drawing light-rays emanating from key points on the object, and finding where these rays are brought to a focus by the mirror. This task can be accomplished using just *four* simple rules:

- An incident ray which is parallel to the principal axis is reflected through the focus F of the mirror.
- An incident ray which passes through the focus F of the mirror is reflected parallel to the principal axis.
- An incident ray which passes through the centre of curvature C of the mirror is reflected back along its own path (since it is normally incident on the mirror).
- An incident ray which strikes the mirror at its vertex V is reflected such that its angle of incidence with respect to the principal axis is equal to its angle of reflection.

The validity of these rules in the paraxial approximation is fairly self-evident.

Consider an object ST which is placed a distance p from a concave spherical mirror. For the sake of definiteness, let us suppose that the object distance p is greater than the focal length f of the mirror. Each point on the object is assumed to radiate light-rays in all directions. Consider four light-rays emanating from the tip T of the object which strike the mirror. The reflected rays are constructed using rules 1-4 above, and the rays are labelled accordingly. It can be seen that the reflected rays all come together at some point T'. Thus, T' is the image of T(*i.e.*, if we were to place a small projection screen at T'then we would see an image of the tip on the screen). As is easily demonstrated, rays emanating from other parts of the object are brought into focus in the vicinity of T' such that a complete image of the object is produced between S' and T' (obviously, point $S'$is the image of point S).

This image could be viewed by projecting it onto a screen placed between points S' and T'. Such an image is termed a *real image*. Note that the image $S'T'$ would also be directly visible to an observer looking straight at the mirror from a distance greater than the image distance q (since the observer's eyes could not tell that the light-rays diverging from the image were in anyway different from those which would emanate from a real object).

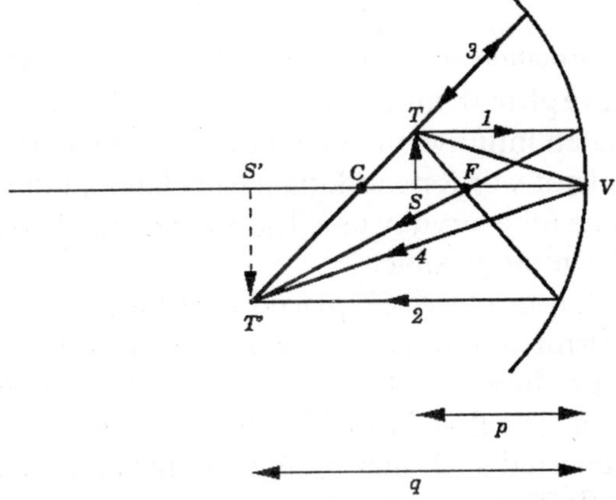

Fig. Formation of a Real Image by a Concave Mirror.

What happens when the object distance p is less than the focal length f. In this case, the image appears to an observer looking straight at the mirror to be located *behind* the mirror. For instance, rays emanating from the tip T of the object appear, after reflection from the mirror, to come from a point T' which is behind the mirror. Note that only two rays are used to locate T', for the sake of clarity. In fact, *two* is the minimum number of rays needed to locate a point image.

Of course, the image behind the mirror cannot be viewed by projecting it onto a screen, because there are no real light-rays behind the mirror. This type of image is termed a *virtual image*. The characteristic difference between a real image and a virtual image is that, immediately after reflection from the mirror, light-rays emitted by the object *converge* on a real image, but *diverge* from a virtual image.

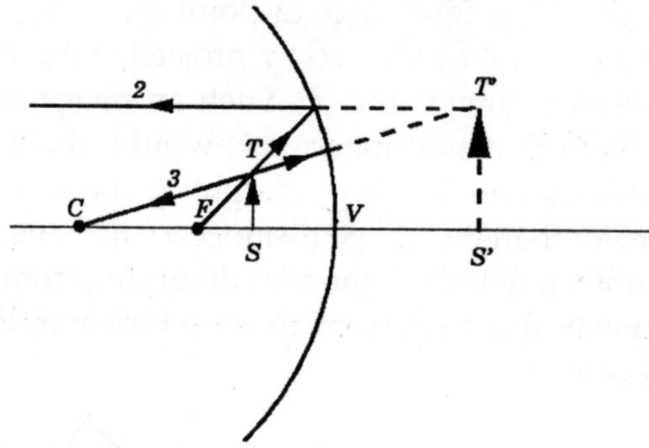

Fig. Formation of a Virtual Image by a Concave Mirror.

The graphical method described above is fine for developing an intuitive understanding of image formation by concave mirrors, or for checking a calculation, but is a bit too cumbersome for everyday use. The analytic method described below is far more flexible.

Consider an object *ST* placed a distance *p* in front of a concave mirror of radius of curvature *R*. In order to find the image *S'T'* produced by the mirror, we draw two rays from *T* to the mirror. The first, labelled 1, travels from *T* to the vertex *V* and is reflected such that its angle of incidence θ equals its angle of reflection.

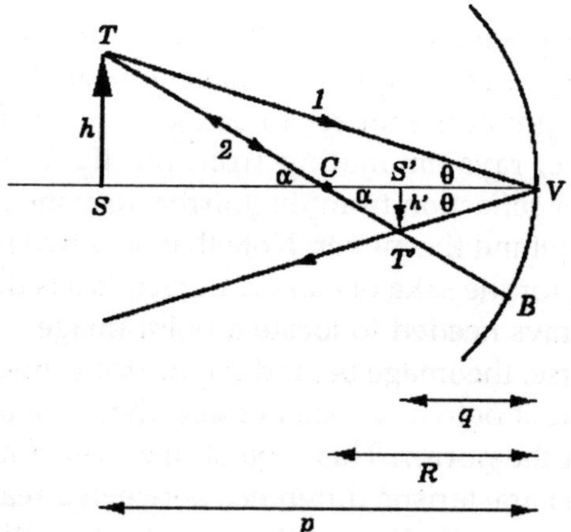

Fig. Image Formation by a Concave Mirror.

The second ray, labelled 2, passes through the centre of curvature C of the mirror, strikes the mirror at point B, and is reflected back along its own path. The two rays meet at point T'. Thus, $S'T'$ is the image of ST, since point S' must lie on the principal axis.

In the triangle STV, we have $\tan\theta = h/p$, and in the triangle $S'T'V$ we have $\tan\theta = -h'/q$, where p is the object distance, and q is the image distance. Here, h is the height of the object, and h' is the height of the image. By convention, h' is a negative number, since the image is inverted (if the image were upright then h' would be a positive number). It follows that

$$\tan\theta = \frac{h}{p} = \frac{-h'}{q}.$$

Thus, the *magnification* M of the image with respect to the object is given by

$$M = \frac{h'}{h} = \frac{q}{p}.$$

By convention, M is negative if the image is inverted with respect to the object, and positive if the image is upright. It is clear that the magnification of the image is just determined by the ratio of the image and object distances from the vertex.

From triangles STC and $S'T'C$, we have $\tan\alpha = h/(p - R)$ and $\tan\alpha = -h'/(R - q)$, respectively. These expressions yield

$$\tan\alpha = \frac{h}{p-R} = -\frac{h'}{R-q}.$$

Equations can be combined to give

$$\frac{-h'}{h} = \frac{R-q}{p-R} = \frac{q}{p},$$

which easily reduces to

$$\frac{1}{p} + \frac{1}{q} = \frac{2}{R}.$$

This expression relates the object distance, the image distance, and the radius of curvature of the mirror.

For an object which is very far away from the mirror (*i.e.*, $p \to \infty$), so that light-rays from the object are parallel to the principal axis, we expect the image to form at the focal point F of the mirror. Thus, in this case, $q = f$, where $f$is the focal length of the mirror, and Eq. reduces to

$$0 + \frac{1}{f} = \frac{2}{R}.$$

The above expression yields

$$f = \frac{R}{2}.$$

In other words, in the paraxial approximation, the focal length of a concave spherical mirror is *half* of its radius of curvature. Equations can be combined to give

$$\frac{1}{p} + \frac{1}{q} = \frac{1}{f} \cdot \delta$$

The above expression was derived for the case of a real image. However, as is easily demonstrated, it also applies to virtual images provided that the following sign convention is adopted. For real images, which always form *in front* of the mirror, the image distance $q$is *positive*. For virtual images, which always form *behind* the mirror, the image distance q is *negative*.

Table Rules for Image Formation by Concave Mirrors

Position of object	*Position of image*	*Character of image*
At ∞	At F	Real, zero size
Between ∞ and C	Between C and ∞	
At C	At C	Real, inverted, diminished
		Real, inverted, same size
Between C and F	Between C and ∞	Real, inverted, magnified
At F	At ∞	
Between F to V	From −∞ to V	Virtual, upright, magnified
At V	At V	Virtual, upright, same size

It immediately follows, from Eq. that real images are always inverted, and virtual images are always upright. The location and character of the image formed in a concave spherical mirror depend on the location of the object, according to Eqs.

It is clear that the *modus operandi* of a shaving mirror, or a makeup mirror, is to place the object (*i.e.*, a face) between the mirror and the focus of the mirror. The image is upright, (apparently) located behind the mirror, and magnified.

Image Formation by Convex Mirrors

The definitions of the principal axis, centre of curvature *C*, radius of curvature *R*, and the vertex *V*, of a convex mirror are analogous to the corresponding definitions for a concave mirror.

When parallel light-rays strike a convex mirror they are reflected such that they appear to emanate from a single point *F* located behind the mirror. This point is called the *virtual focus* of the mirror. The focal length *f* of the mirror is simply the distance between *V* and *F*. As is easily demonstrated, in the paraxial approximation, the focal length of a convex mirror is half of its radius of curvature.

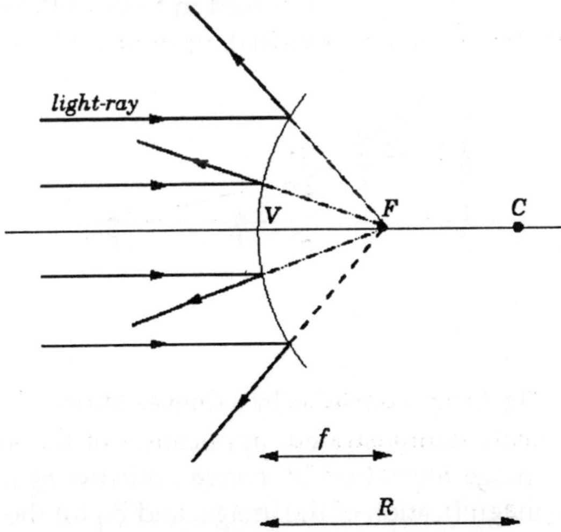

Fig. The Virtual Focus of a Convex Mirror.

There are, again, two alternative methods of locating the image formed by a convex mirror. The first is graphical, and the second analytical.

According to the graphical method, the image produced by a convex mirror can always be located by drawing a ray diagram according to *four* simple rules:

- An incident ray which is parallel to the principal axis is reflected as if it came from the virtual focus *F* of the mirror.
- An incident ray which is directed towards the virtual focus *F* of the mirror is reflected parallel to the principal axis.
- An incident ray which is directed towards the centre of curvature *C* of the mirror is reflected back along its own path (since it is normally incident on the mirror).
- An incident ray which strikes the mirror at its vertex *V* is reflected such that its angle of incidence with respect to the principal axis is equal to its angle of reflection.

The validity of these rules in the paraxial approximation is, again, fairly self-evident. The two rays are used to locate the image *S'T'* of an object *ST* placed in front of the mirror. It can be seen that the image is virtual, upright, and diminished.

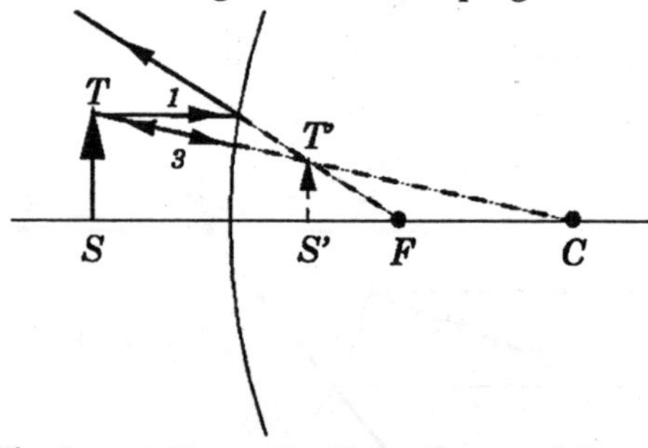

Fig. Image Formation by a Convex Mirror.

As is easily demonstrated, application of the analytical method to image formation by convex mirrors again yields Eq. for the magnification of the image, and Eq for the location of the image, provided that we adopt the following sign convention.

The focal length *f* of a convex mirror is redefined to be *minus* the distance between points *V* and *F*. In other words, the focal length of a concave mirror, with a real focus, is always positive, and the focal length of a convex mirror, with a virtual focus, is always negative.

The location and character of the image formed in a convex spherical mirror depend on the location of the object, according to Eqs. (with $f < 0$).

Table Rules for Image Formation by Convex Mirrors

Position of object	*Position of image*	*Character of Image*
At ∞	At *F*	Virtual, zero size
Between ∞ and V	Between *F* and *V*	Virtual, upright, diminished
At *V*	At *V*	Virtual, upright, same size

In summary, the formation of an image by a spherical mirror involves the *crossing* of light-rays emitted by the object and reflected off the mirror. If the light-rays actually cross in front of the mirror then the image is real. If the light-rays do not actually cross, but appear to cross when projected backwards behind the mirror, then the image is virtual. A real image can be projected onto a screen, a virtual image cannot. However, both types of images can be seen by an observer, and photographed by a camera.

The magnification of the image is specified by Eq., and the location of the image is determined by Eq.. These two formulae can be used to characterize both real and virtual images formed by either concave or convex mirrors, provided that the following sign conventions are observed:

- The height h' of the image is positive if the image is upright, with respect to the object, and negative if the image is inverted.
- The magnification M of the image is positive if the image is upright, with respect to the object, and negative if the image is inverted.
- The image distance q is positive if the image is real, and, therefore, located in front of the mirror, and negative if the image is virtual, and, therefore, located behind the mirror.
- The focal length f of the mirror is positive if the mirror is concave, so that the focal point F is located

in front of the mirror, and negative if the mirror is convex, so that the focal point F is located behind the mirror.

Note that the front side of the mirror is defined to be the side from which the light is incident.

IMAGE FORMATION BY PLANE MIRRORS

Both concave and convex spherical mirrors asymptote to plane mirrors in the limit in which their radii of curvature R tend to infinity. In other words, a plane mirror can be treated as either a concave or a convex mirror for which $R \to \infty$. Now, if $R \to \infty$, then f = ± $R/2 \to \infty$, so $1/f \to 0$, and Eq. yields

$$\frac{1}{p} + \frac{1}{q} = \frac{1}{f} = 0,$$

or

$$q = -p.$$

Thus, for a plane mirror the image is *virtual*, and is located as far behind the mirror as the object is in front of the mirror. According to Eq. the magnification of the image is given by

$$m = -\frac{q}{p} = 1.$$

Clearly, the image is *upright*, since $M > 0$, and is the *same* size as the object, since $|M| = 1$. However, an image seen in a plane mirror does differ from the original object in one important respect: *i.e.*, left and right are *swapped over*. In other words, a right-hand looks like a left-hand in a plane mirror, and *vice versa*.

THIN LENSES

A lens is a transparent medium (usually glass) bounded by two curved surfaces (generally either spherical, cylindrical, or plane surfaces).

The line which passes normally through both bounding surfaces of a lens is called the *optic axis*. The point O on the optic axis which lies midway between the two bounding surfaces is called the *optic centre*.

There are two basic kinds of lenses: *converging*, and *diverging*. A converging lens brings all incident light-rays parallel to its optic axis together at a point F, behind the lens, called the *focal point*, or *focus*, of the lens.

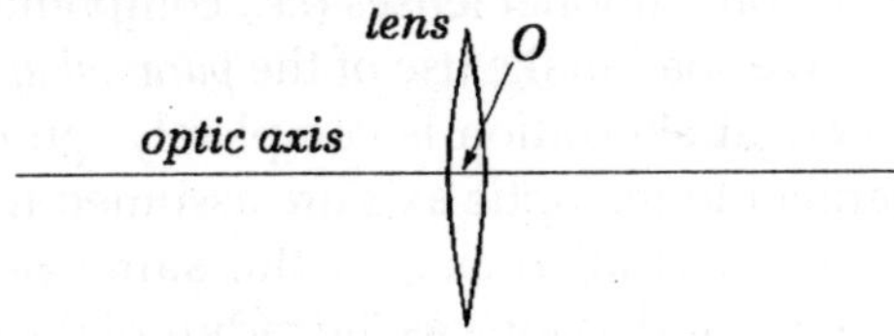

Fig. The Optic Axis of a Lens.

A diverging lens spreads out all incident light-rays parallel to its optic axis so that they appear to diverge from a *virtual focal point* F in front of the lens. Here, the front side of the lens is conventionally defined to be the side from which the light is incident. The differing effects of a converging and a diverging lens on incident light-rays parallel to the optic axis (*i.e.*, emanating from a distant object).

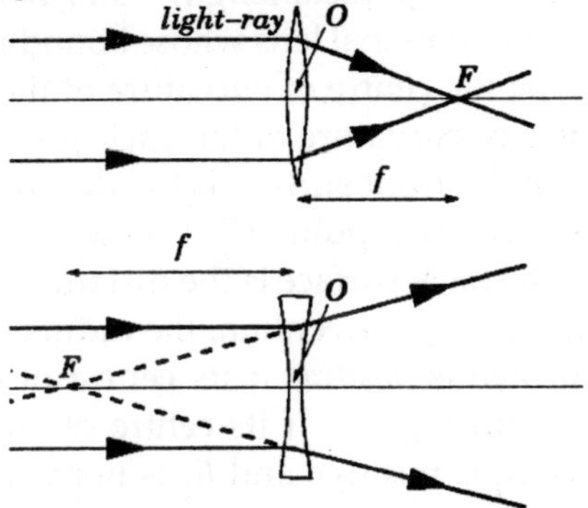

Fig. The Focii of Converging (Top) and Diverging (Bottom) Lens.

Lenses, like mirrors, suffer from *spherical aberration*, which causes light-rays parallel to the optic axis, but a relatively long way from the axis, to be brought to a focus, or a virtual focus, *closer* to the lens than light-rays which are relatively close to the axis.

It turns out that spherical aberration in lenses can be completely cured by using lenses whose bounding surfaces are *non-spherical*. However, such lenses are more difficult, and, therefore, more expensive, to manufacture than conventional lenses whose bounding surfaces are spherical. Thus, the former sort of lens is only employed in situations where the spherical aberration of a conventional lens would be a serious problem.

The usual method of curing spherical aberration is to use *combinations* of conventional lenses (*i.e.*, compound lenses). In the following, we shall make use of the *paraxial approximation*, in which spherical aberration is completely ignored, and all light-rays parallel to the optic axis are assumed to be brought to a focus, or a virtual focus, at the same point F. This approximation is valid as long as the radius of the lens is small compared to the object distance and the image distance.

The *focal length* of a lens, which is usually denoted f, is defined as the distance between the optic centre O and the focal point F. However, by convention, *converging* lenses have *positive* focal lengths, and *diverging* lenses have *negative* focal lengths. In other words, if the focal point lies behind the lens then the focal length is positive, and if the focal point lies in front of the lens then the focal length is negative.

Consider a conventional lens whose bounding surfaces are *spherical*. Let C_f be the centre of curvature of the front surface, and C_b the centre of curvature of the back surface. The radius of curvature R_f of the front surface is the distance between the optic centre O and the point C_f. Likewise, the radius of curvature R_b of the back surface is the distance between points O and C_b. However, by convention, the radius of curvature of a bounding surface is *positive* if its centre of curvature lies *behind* the lens, and *negative* if its centre of curvature lies *in front* of the lens. R_f is positive and R_b is negative.

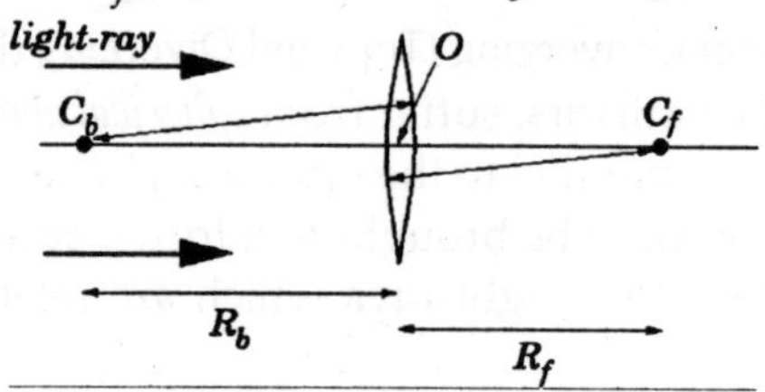

Fig. A Thin Lens.

In the paraxial approximation, it is possible to find a simple formula relating the focal length f of a lens to the radii of curvature, R_f and R_b, of its front and back bounding surfaces. This formula is written

$$\frac{1}{f} = (n-1)\left(\frac{1}{R_f} - \frac{1}{R_b}\right),$$

where n is the refractive index of the lens. The above formula is usually called the *lens-maker's formula,* and was discovered by Descartes. Note that the lens-maker's formula is only valid for a *thin lens* whose thickness is small compared to its focal length. What Eq. is basically telling us is that light-rays which pass from air to glass through a *convex* surface are *focused,* whereas light-rays which pass from air to glass through a *concave* surface are *defocused.*

Furthermore, since light-rays are *reversible,* it follows that rays which pass from glass to air through a *convex* surface are *defocused,* whereas rays which pass from air to glass through a *concave* surface are *focused.*

Note that the net focusing or defocusing action of a lens is due to the *difference* in the radii of curvature of its two bounding surfaces.

Suppose that a certain lens has a focal length f. What happens to the focal length if we turn the lens around, so that its front bounding surface becomes its back bounding surface, and *vice versa*? It is easily seen that when the lens is turned around $R_f \to -R_b$ and $R_b \to -R_f$. However, the focal length f of the lens is invariant under this transformation, according to Eq.

Thus, the focal length of a lens is the same for light incident from either side. In particular, a converging lens remains a converging lens when it is turned around, and likewise for a diverging lens.

The most commonly occurring type of converging lens is a *bi-convex,* or *double-convex,* lens, for which $R_f > 0$ and $R_b < 0$. In this type of lens, both bounding surfaces have a focusing effect on light-rays passing through the lens. Another fairly common type of converging lens is a *plano-convex* lens, for which $R_f > 0$ and $R_b = \infty$. In this type of lens, only the curved bounding surface has a focusing effect on light-rays. The plane surface has no focusing or defocusing effect. A less common type of converging lens is a *convex-meniscus* lens, for which $R_f > 0$ and $R_b > 0$, with $R_f < R_b$. In this type of lens, the front bounding surface has a focusing effect on light-rays, whereas the back bounding surface has a defocusing effect, but the focusing effect of the front surface wins out.

The most commonly occurring type of diverging lens is a *bi-concave,* or *double-concave,* lens, for which $R_f < 0$ and $R_b > 0$. In this type of lens, both bounding surfaces have a defocusing effect on light-rays passing through the lens.

Another fairly common type of converging lens is a *plano-concave* lens, for which $R_f < 0$ and $R_b = \infty$. In this type of lens, only the curved bounding surface has a defocusing effect on light-rays.

The plane surface has no focusing or defocusing effect. A less common type of converging lens is a *concave-meniscus* lens, for which $R_f < 0$ and $R_b < 0$, with $R_f < |R_b|$. In this type of lens, the front bounding surface has a defocusing effect on light-rays, whereas the back bounding surface has a focusing effect, but the defocusing effect of the front surface wins out.

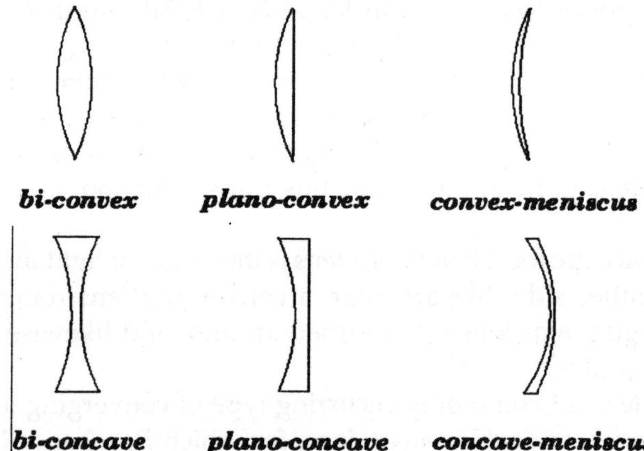

Fig. Various Different types of Thin Lens.

The various types of lenses mentioned above. Note that, as a general rule, converging lenses are thicker at the centre than at the edges, whereas diverging lenses are thicker at the edges than at the centre.

Image Formation by Thin Lenses

There are two alternative methods of locating the image formed by a thin lens. Just as for spherical mirrors, the first method is *graphical*, and the second *analytical*.

The graphical method of locating the image formed by a thin lens involves drawing light-rays emanating from key points on the object, and finding where these rays are brought to a focus by the lens. This task can be accomplished using a small number of simple rules.

Consider a converging lens. It is helpful to define *two* focal points for such a lens. The first, the so-called *image focus*, denoted F_i, is defined as the point behind the lens to which all incident light-rays parallel to the optic axis converge after passing through the lens. This is the same as the focal point F defined previously. The second, the so-called *object focus*, denoted F_a, is defined as the position in front of the lens for which rays emitted from a point source of light placed at that position would be refracted parallel to the optic axis after passing through the lens. It is easily demonstrated that the object focus F_o is as far in front of the optic centre O of the lens as the image focus F_i is behind O. The distance from the optic centre to either focus is, of course, equal to the focal length f of the lens. The image produced by a converging lens can be located using just *three* simple rules:

- An incident ray which is parallel to the optic axis is refracted through the image focus F_i of the lens.
- An incident ray which passes through the object focus F_o of the lens is refracted parallel to the optic axis.
- An incident ray which passes through the optic centre O of the lens is not refracted at all.

The last rule is only an approximation. It turns out that although a light-ray which passes through the optic centre of the lens does not change direction, it is displaced slightly to one side. However, this displacement is negligible for a thin lens.

The image $S'T'$of an object ST placed in front of a converging lens is located using the above rules. In fact, the three rays, 1-3, emanating from the tip T of the object, are constructed using rules 1-3, respectively. Note that the image is real (since light-rays actually cross), inverted, and diminished.

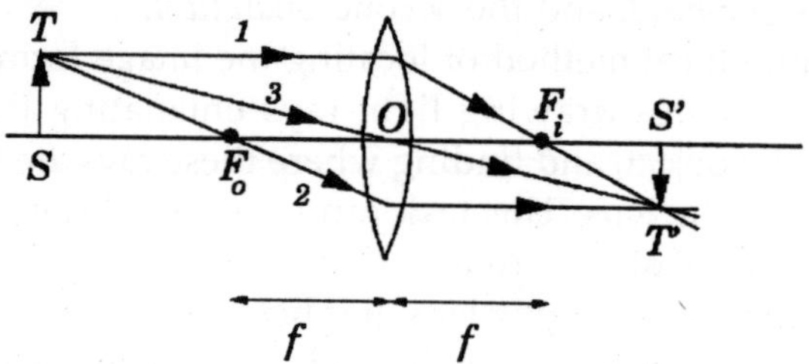

Fig. Image Formation by A Converging Lens.

Consider a diverging lens. It is again helpful to define two focal points for such a lens. The image focus F_i is defined as the point in front of the lens from which all incident light-rays parallel to the optic axis appear to diverge after passing through the lens. This is the same as the focal point F defined earlier. The object focus F_o is defined as the point behind the lens to which all incident light-rays which are refracted parallel to the optic axis after passing through the lens appear to converge. Both foci are located a distance f from the optic centre, where f is the focal length of the lens. The image produced by a diverging lens can be located using the following three rules:

- An incident ray which is parallel to the optic axis is refracted as if it came from the image focus F_i of the lens.
- An incident ray which is directed towards the object focus F_o of the lens is refracted parallel to the optic axis.
- An incident ray which passes through the optic centre O of the lens is not refracted at all.

The image $S'T'$of an object ST placed in front of a diverging lens is located using the above rules. In fact, the three rays, 1–3, emanating from the tip $T$of the object, are constructed using rules 1-3, respectively. Note that the image is virtual (since light-rays do not actually cross), upright, and diminished.

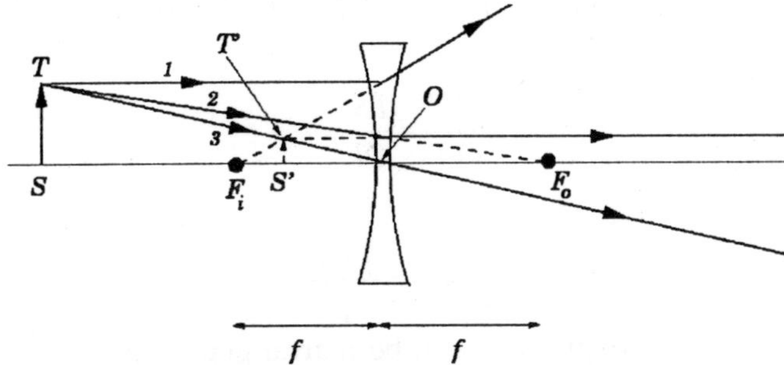

Fig. Image Formation by A Diverging Lens.

Let us now investigate the analytical method. Consider an object of height h placed a distance p in front of a converging lens. Suppose that a real image of height h' is formed a distance q behind the lens.

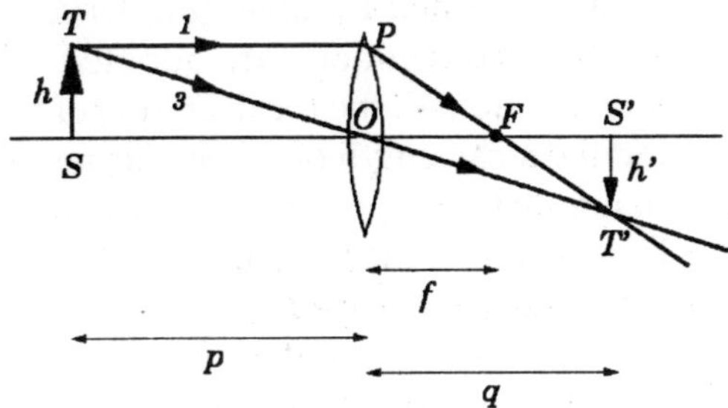

Fig. Image Formation by A Converging Lens.

Now, the right-angled triangles SOT and $S'OT'$ are similar, so

$$\frac{-h'}{h} = \frac{OS'}{OS} = \frac{q}{p}.$$

Here, we have adopted the convention that the image height h' is *negative* if the image is *inverted*. The magnification of a thin converging lens is given by

$$M = \frac{-h'}{h} = -\frac{q}{p}.$$

This is the same as the expression for the magnification of a spherical mirror. Note that we are again adopting the convention that the magnification is *negative* if the image is *inverted*.

The right-angled triangles $O\,P\,F$ and $S'T'F$ are also similar, and so

$$\frac{S'T'}{O\,P}=\frac{FS'}{OF},$$

or

$$\frac{-h'}{h}=\frac{q}{p}=\frac{q-f}{f}.$$

The above expression can be rearranged to give

$$\frac{q}{p}+\frac{1}{q}=\frac{1}{f}.$$

Although formulae were derived for the case of a real image formed by a converging lens, they also apply to virtual images, and to images formed by diverging lenses, provided that the following sign conventions are adopted. First of all, as we have already mentioned, the focal length f of a *converging* lens is *positive,* and the focal length of a *diverging* lens is *negative.* Secondly, the image distance q is *positive* if the image is *real,* and, therefore, located *behind* the lens, and *negative* if the image is *virtual,* and, therefore, located *in front* of the lens. It immediately follows, from Eq. that *real* images are always *inverted,* and *virtual* images are always *upright.*

The location and character of the image formed by a converging lens depend on the location of the object. Here, the point V_0 is located on the optic axis two focal lengths in front of the optic centre, and the point V_i is located on the optic axis two focal lengths behind the optic centre. Note the almost exact analogy between the image forming properties of a converging lens and those of a concave spherical mirror.

Table. Rules For Image Formation by Converging Lenses

Position of Object	*Position of Image*	*Character of Image*
At $+\infty$	At F	Real, zero size
Between $+\infty$ and V_0	Between F and V_i	Real, inverted, diminished
At V_0	At V_i	Real, inverted, same size
Between V_0 and F	Between V_i and $-\infty$	Real, inverted, magnified
At F	At $-\infty$	
Between F and O	From $+\infty$ to O	Virtual, upright, magnified
At O	At O	Virtual, upright, same size

The location and character of the image formed by a diverging lens depend on the location of the object. Note the almost exact analogy between the image forming properties of a diverging lens and those of a convex spherical mirror.

Table. Rules for Image Formation by Diverging Lenses

Position of Object	*Position of Image*	*Character of Image*
At ∞	At F_i	Virtual, zero size
Between ∞ and O	Between F_i and O	Virtual, upright, iminished
At O	At O	Virtual, upright, same size

Finally, let us reiterate the sign conventions used to determine the positions and characters of the images formed by thin lenses:

- The height h' of the image is positive if the image is upright, with respect to the object, and negative if the image is inverted.
- The magnification M of the image is positive if the image is upright, with respect to the object, and negative if the image is inverted.
- The image distance q is positive if the image is real, and, therefore, located behind the lens, and negative if the image is virtual, and, therefore, located in front of the lens.
- The focal length f of the lens is positive if the lens is converging, so that the image focus F_i is located behind the lens, and negative if the lens is diverging, so that the image focus F_i is located in front of the lens.

Note that the front side of the lens is defined to be the side from which the light is incident.

Chromatic aberration

We have seen that both mirrors and lenses suffer from spherical aberration, an effect which limits the clarity and sharpness of the images formed by such devices. However, lenses also suffer from another type of abberation called *chromatic abberation*. This occurs because the index of refraction of the glass in a lens is different for different wavelengths. We have seen that a prism refracts violet light more than red light.

The same is true of lenses. As a result, a simple lens focuses violet light closer to the lens than it focuses red light. Hence, white light produces a slightly blurred image of an object, with coloured edges.

For many years, chromatic abberation was a sufficiently serious problem for lenses that scientists tried to find ways of reducing the number of lenses in scientific instruments, or even eliminating them all together. For instance, Isaac Newton developed a type of telescope, now called the Newtonian telescope, which uses a mirror instead of a lens to collect light. However, in 1758, John Dollond, an English optician, discovered a way to eliminate chromatic abberation. He combined two lenses, one converging, the other diverging, to make an *achromatic doublet*.

The two lenses in an achromatic doublet are made of different type of glass with indices of refraction chosen such that the combination brings any two chosen colours to the same sharp focus. Modern scientific instruments use *compound lenses* (*i.e.*, combinations of simple lenses) to simultaneously eliminate both chromatic and spherical aberration.

Chapter 6

Wave Generation and Propagation

To perform realistic simulations of green water and bow impact loading, the method has to be extended with wave generation options. There are different ways to create a wave field in a computational program:

- Prescribe a wave at the inflow boundary using theory, for example Stokes theory or a superposition of linear waves.
- Use a wave maker, which can be modelled as a moving object.
- Use another calculation method, which calculates kinematics away from an object that will be prescribed at the domain boundaries.

The waves considered are long crested waves, so the incoming wave is two-dimensional of shape. Of course, when an object is present, the wave will be disturbed and three-dimensional aspects become important. At the boundary of the domains appropriate conditions are needed to let the wave travel in and out of the domain in an undisturbed way. Therefore, at the inflow boundary fluid wave kinematics will be prescribed according to wave description theories.

The undisturbed wave will be prescribed, so the disturbance due to wave diffraction on an object is not taken into account. At the outflow boundary conditions should be imposed, such that the wave can leave the domain undisturbedly. Determining these conditions is difficult, because no information about the wave is present near the outflow boundary. After describing the wave theories and conditions at the outflow boundary, attention will be paid to possible energy dissipation due to the upwind discretisation of the convective terms.

Also the influence of the treatment of the free surface on wave propagation, like the choice for free surface velocities and the displacement algorithm, is investigated. To validate the numerical model, simulations have been performed, which can be checked against theory or experiments. First, results will be shown of two-dimensional wave simulations without the presence of an object. Further, a spar platform has been put into the flow.

Forces and wave heights have been calculated, which can be compared with experimental results. Finally, as the most demanding test case, green water on the bow of a moving FPSO is simulated.

WAVE

At the inflow boundary the wave will be initiated using a theoretical wave description. In this thesis the wave is usually travelling in positive x-direction, making the left domain wall the inflow boundary. The wave is prescribed as head wave, but of course by rotating the body in the domain every wave direction can be achieved. For the simulation of a wave in the domain, the velocities and wave height given by wave theory are prescribed.

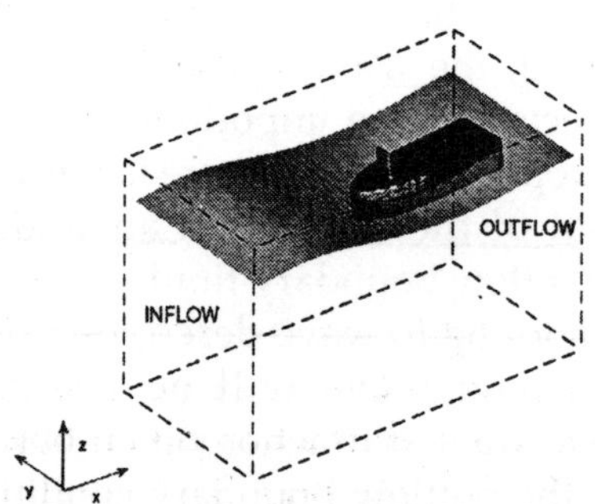

Fig. Coordinate System Used in the Simulation: The Wave Travels in Positive Xdirection

The waves that can be formed in COM FLOW are based on Airy wave theory describing linear waves, and 5th order Stokes theory describing nonlinear waves. Also a superposition of linear wave components can be used to generate an irregular wave.

Airy Wave Theory

The theory for the generation of a linear wave can be derived in the following way. Mass conservation in a volume filled with an incompressible fluid leads to the continuity equation

$$\nabla . u = 0,$$

with u = (u,w) the velocity vector in two dimensions. The velocity potential F is defined by the equations

$$\frac{\partial \Phi}{\partial x} = u, \frac{\partial \Phi}{\partial z} = w,$$

with x and z the horizontal and vertical coordinate, respectively. Substituting F into the continuity equation leads to a Laplace equation for the velocity potential.

$$\frac{\partial^2 \Phi}{\partial x^2} + \frac{\partial^2 \Phi}{\partial z^2} = 0.$$

When assuming the water surface slope very small, the potential is written as a sine with an amplitude depending on the water depth

$$\Phi(x, z, t) = P(z)\sin(\omega t - kx + \phi),$$

where t is the time, k the wave number, and ϕ the phase angle. P(z) can be determined by substituting Equation into the Laplace equation and solving the differential equation for P(z), which finally results in

$$\Phi(x, z, t) = (C_1 e^{kz} + C_2 e^{-k2})\sin(\omega t - kx + \phi),$$

The constants C1 and C2 can be determined using the boundary conditions at the sea bed and the free surface. At the sea bed a no-leak condition holds, which is given by

$$\frac{\partial \Phi}{\partial z} = 0 \quad \text{at } z = -h.$$

Substituting this boundary condition into Equation reduces the two unknowns to one:

$$\Phi(x,z,t) = C\cosh(k(z+h)\sin(\omega t - kx + \phi).$$

The unknown constant C can be determined by using the free surface dynamic boundary condition that is deduced from the Bernoulli equation, taking into account the small wave steepness.

The condition at the free surface is linearised around the calm water level (z = 0), which leads to

$$\frac{\partial\Phi}{\partial t} + g\zeta = 0 \quad \text{for } z = 0.$$

The free surface elevation ζ can be derived from this equation after substitution of the velocity potential, resulting in

$$\zeta(x,t) = \zeta_a\cos(\omega t - kx) \text{ with } \zeta_a = -\frac{\omega}{g}C\cosh(kh).$$

Rewriting the expression for ζ_a, the constant C can be determined as $C = -\frac{\zeta_a g}{\omega} - \frac{1}{\cosh(kh)}$. Substituting this into Eq leads to the final expression for the velocity potential

$$\Phi(x,z,t) = -\frac{\zeta_a g}{\omega}\frac{\cosh(k(z+h))}{\cosh(kh)}\sin(\omega t - kx + \phi).$$

The linearised free surface kinematic boundary condition given by

$$\frac{\partial z}{\partial t} + \frac{1}{g}\frac{\partial^2\Phi}{\partial t^2} = 0 \text{ for } z = 0$$

leads to the dispersion relation, which connects the wave number and frequency in the following way:

$$k\tanh(kh) = \frac{\omega^2}{g}.$$

For deep water, the dispersion relation reduces to $k = \omega^2/g$.

$$u(t,x,z) = \frac{\zeta_a g k}{\omega}\cos(\omega t - kx + \phi)\frac{\cosh(k(z+h))}{\cosh(kh)},$$

$$u(t,x,z) = \frac{\zeta_a g k}{\omega} \sin(\omega t - kx + \phi) \frac{\cosh(k(z+h))}{\cosh(kh)}.$$

As stated before, linear theory can only be applied to very low waves. The range of suitability of linear theory in deep water is $H/\lambda < 0.0062$, with H the wave height and λ the wavelength. Linear wave theory can also be used to generate irregular seas, because the superposition principle can be applied. Given vectors of frequencies, amplitudes and phases, a linear wave can be built as the sum of the individual components.

Wave Kinematics Above the Calm Water Level

Since linear wave theory is only valid up to the calm water level and velocities are needed up to the free surface, some kind of stretching technique has to be used above the calm water level. In this method Wheeler stretching has been used as a commonly accepted method, which is easy to implement. In the Wheeler stretching technique the negative z-axis has been extended from the actual instantaneous free surface elevation to the sea bed. This has been done by replacing z in the right-hand-side of Equations by

$$z' = \frac{h}{h+\zeta} z + h\left(\frac{h}{h+\zeta} - 1\right)$$

where

- z' is the computational vertical coordinate $-h \le z' \le 0$;
- z is the actual vertical coordinate $- \le z \le \zeta$.

5th order Stokes Theory

For steeper waves, where in general the crests become higher and the troughs flatter, linear theory does no longer hold. To describe nonlinear waves, a solution to potential theory is used.

The solution is represented by Fourier series, and the coefficients in these series can be written as perturbation expansions with parameter Ak. Here, k is the wave number, which can be written in terms of the wave length $k - 2\pi/\lambda$, and A is the amplitude of the wave at lowest order. The terms in the perturbation expansion can be found by satisfying boundary conditions on the free surface, and solving the resulting set of ordered equations. The expansion of the series can be done, in theory, infinitely far, but in practice, the 5th order solution is already very complicated. Details about how to implement the 5th order Stokes solution can be found in , where the sign correction described by should be taken into account.

TREATMENT OF OPEN BOUNDARIES

Domain walls in wave simulations at open sea need to permit fluid flow in and out. Therefore, at least an inflow boundary is needed where the wave is generated, and an outflow boundary opposite to the inflow boundary where the wave leaves the domain. In the current method the waves, which are long crested, usually travel in the positive x-direction.

Because of the two-dimensional shape, the side boundaries can often be closed walls; when positioned far enough from the structure they do not influence the wave field in the close surroundings of the structure. When the waves are highly distorted, such that side walls are severely influencing the flow, also outflow boundary conditions can be used at the side walls a cross-section of one row of cells is shown with the positions of the velocities at the inflow and outflow boundaries. At the inflow boundary the velocities and VOF function in the column of cells directly left of the domain boundary are prescribed.

This way, the horizontal inflow velocity is positioned at the domain boundary, and no pressure is needed in the inflow cells. For the outflow boundary the horizontal velocity is shifted to one cell width away of the actual domain boundary. Therefore, also a pressure is needed in the outflow cells, so velocities, pressure and VOF function need to be determined at the outflow boundary (using a Neumanntype outflow condition). The wave descriptions have been given, from which velocities and wave height at the inflow boundary are derived. The remainder of this section deals with the outflow boundary.

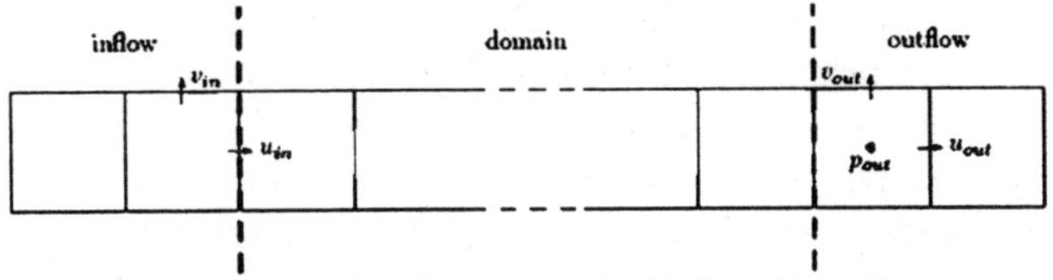

Fig. Position of Velocities at the Inflow Boundary, and Velocities and Pressure at the Outflow Boundary

Overview Outflow Boundary Conditions

A very important aspect of wave simulation is to determine the conditions at the outflow boundaries. If the wave is developing inside the domain, it should flow out of the domain as if there was no boundary. Otherwise, the wave will reflect from the boundaries into the domain, which disturbs the simulation. An overview of outflow boundary conditions that prevent wave reflections is given by Givoli .

- The first class consists of special procedures for the numerical solution of wave problems in unbounded domains, that involve an artificial boundary but not the direct use of a non-reflecting boundary condition. Cerjan and others presented what can be termed a 'filtering scheme'. In this scheme, the amplitudes of the displacements are gradually reduced in a strip of nodes adjacent to the boundary. Thus, the solution is artificially damped in the vicinity of the boundary. Another kind of dissipation zone was used by where an extra damping pressure was added to the free surface, which opposes the vertical wave velocity. A disadvantage of such damping zones is the increase of computational cells out of which the damping zone exists. Especially, in three dimensions many computational cells have to be added outside the real computational domain.
- In the second class, which is the largest one, local non-reflecting boundary conditions (NRBCs) based on the scalar wave equation are used. There exist

many variations in local NRBCs, every problem in different types of fields uses NRBCs that works best for that particular problem. A comparison of different boundary conditions derived from the discretisation of the multi-dimensional wave. The widely used NRBC of Sommerfeld is a discretisation of the one-dimensional scalar wave equation with an a priori chosen wave velocity. A variation hereof was introduced by Orlanski, who calculated the wave velocity for every grid point and used that velocity in the discretised wave equation. The advantage of this method is that no information is needed of the flow beforehand.

- Finally, non-reflecting boundary conditions are considered that are non-local in time or space or both. These NRBCs have the disadvantage that many time levels should be stored in memory.

The Sommerfeld and Orlanski boundary conditions is given. The implementation of a general non-reflecting boundary condition is described, and the damping zone used by is introduced.

Non-Reflecting Boundary Conditions

Boundary conditions based on the wave equation One way to prevent waves reflecting from the outflow boundary is to use a non-reflecting boundary condition. In case of waves most of the existing boundary conditions are based on the wave equation

$$\frac{\partial \phi}{\partial t} + c\frac{\partial \phi}{\partial x} = 0,$$

where ϕ is any quantity that travels wave-like as velocity or pressure, and c the wave velocity. In Equation the direction of the wave is in the x-direction. In the Sommerfeld boundary condition the wave velocity is chosen a priori, which is easy when regular waves are simulated. In that case, the Sommerfeld condition gives very good results as will be shown later. When irregular waves are considered, or when the regular waves are deformed due to the presence of an object, the method does not give very accurate results, since only one wave velocity can be chosen.

For those situations, Orlanski developed a method where the wave velocity is not chosen a priori, but is calculated every time step based on the local wave kinematics near the outflow boundary. The problem with this method is to determine the wave velocity accurately.

To calculate c, first the wave equation is discretised using finite differences in cells close to the outflow boundary where the condition will be imposed.

The wave equation is discretised for the horizontal velocity in every level of cells in the water depth. The c is then calculated as the mean of the approximated wave velocities in every level in the water depth.

The calculated wave velocity is shown for a regular wave with an actual wave velocity of 22.6 m/s during one wave period. The calculated wave velocity is not constant at all, but oscillates around the theoretical value. The jump that occurs around 7.2 s is present because there the crest of the wave travels through the outflow boundary resulting in $\partial u/\partial = 0$. This can cause the wave velocity to jump from minus infinity to plus infinity.

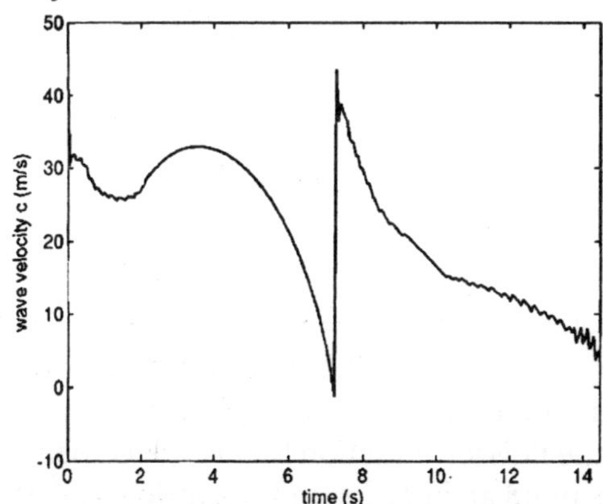

Fig. Calculated Wave Velocity Using a Finite Difference Approximation of the Wave Equation

In this region, the wave velocity is adapted, such that these extreme values are not used in the outflow boundary conditions. In an investigation of wave propagation using Orlanski's method at the outflow boundary, it turned out that the results were not very accurate, so this method is not used in the simulations shown in this thesis.

The Implementation of a Non-Reflecting Boundary Condition

At the outflow boundary conditions for pressure and velocity are needed. When using a Sommerfeld boundary condition, for the velocities at the outflow boundary, the wave equation Equation is discretised using finite differences in the following way

$$\frac{\phi_e^{n+1} - \phi_e^n}{\delta t} + C\frac{\phi_e^n - \phi_c^n}{\delta x_e} = 0,$$

and ϕ is used for the velocity components. From this equation the velocities at the outflow boundary u_e^{n+1}, v_e^{n+1} and w_e^{n+1} follow.

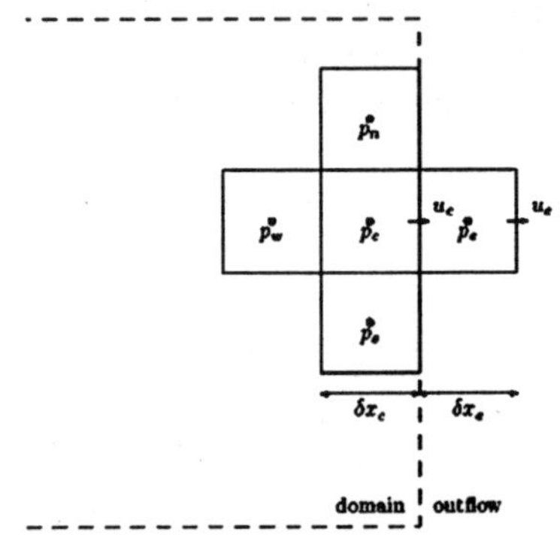

Fig. Configuration of Cells Near the Outflow Boundary

For the calculation of the pressure in the outflow cell the procedure is somewhat more complicated, since the pressure appears in the Poisson equation that is solved using an iterative method. To avoid an iteration in the outflow cell, the boundary condition is substituted in the equation for the interior pressure cell pe, such that the equation for the pressure in the outflow cell is not needed. At the end of the iterations the pressure in the outflow cell is updated using the pressure in the interior cell. This procedure is elaborated for a general outflow boundary condition

$$\alpha \frac{p_e - p_c}{\delta x_{pe}} + \beta p_e = \alpha A + \beta p_0,$$

Here, the Neumann condition can recognised for (α, β) = (1, 0) and the Dirichlet condition for (α, β) = (0, 1). The Sommerfeld condition is obtained when taking $\alpha = c$, $\beta = 1/\delta t$, $A = 0$ and $p_0 = p_e^n$. This equation is discretised using finite differences

$$\alpha \frac{p_e - p_c}{\delta x_{pe}} + \beta p_e = \alpha A + \beta p_0,$$

with $\delta x_{pe} = (\delta x_e + \delta x_c)/2$. From this equation the pressure in the outflow cell pe can be solved

$$p_e = \frac{\alpha}{\alpha + \beta \delta x_{pe}} p_c + \frac{\alpha A \delta x_{pe} + \beta p_0 \delta x_{pe}}{\alpha + \beta \delta x_{pe}}.$$

$$C_c p_c + C_w p_w + C_e p_e + C_n p_n + C_s p_s = RHS.$$

The coefficients Cc to Cs contain squared grid sizes and the right-hand-side RHS contains the divergence of the contributions from convection, diffusion and external forces. In this equation the outflow pressure given by Equation can be substituted resulting in

$$\left(C_c \frac{\alpha}{\alpha + \beta \delta x_{pe}} C_e \right) p_c + C_w p_w + C_n p_n + C_s p_s$$

$$= RHS - \frac{\alpha A \delta x_{pe} + \beta p_0 \delta x_{pe}}{\alpha + \beta \delta x_{pe}} C_e.$$

The matrix containing the coefficients remains diagonal dominant when α, β > 0. After the pressure in the interior of the domain is solved from the Poisson equation, the pressure in the outflow cell is updated using Equation.

Pressure Damping at the Free Surface

Instead of using a non-reflecting boundary condition, a dissipation zone can be used where the wave is damped. In the current method, for the damping inside the dissipation zone a pressure term has been added to the free surface pressure. Physically, this can be interpreted as the air acting like a damper on the wave. The pressure added to the atmospheric pressure term at the free surface is chosen as a function of the vertical velocity at the free surface:

$$pdamp(t,x,\zeta) = \alpha(x)\, w(t,x,\zeta).$$

The damping function $\alpha(x)$ should be chosen such that the wave is damped completely, and the wave should not reflect at the start of the damping zone. A polynomial form for the function is used by concludes in his report that a linear function is suitable for the purposes of wave simulation. If a linear damping function $\alpha(x) = ax + b$ is used, two constants a and b have to be chosen. Here, a is the slope of the damping function, which determines the rate of the damping. The constant b has to be chosen such that the damping function is zero at the start of the dissipation zone.

The slope of the damping function is determined by the characteristics of the wave that is to be damped. Meskers shows in his report how the slope and the length of the dissipation zone can be determined after choosing the total reflection that is allowed. For a number of allowed reflection factors the slope and length of the dissipation zone are plotted. For regular waves, the following steps have to be taken to determine the slope and length of the dissipation zone. First, choose the amount of reflection (rtot) that can be permitted. This amount of reflection is calculated theoretically, and is only valid for a perfectly regular wave.

Then, given the wave frequency and reflection coefficient, determine the length of the numerical beach using the curved lines. At the left coordinate axis, the accompanying values are placed. Finally, determine the slope of the numerical beach, using the straight lines with the values of the right coordinate axis. Besides the choice of the function , there also are some different possibilities for the closing wall of the domain, by which we mean the wall opposite of the inflow boundary.

- In the first option the wave is damped completely, the closing wall of the domain is a solid wall, no water is flowing out.

- In the second option the wave is damped towards its analytical form. So a 'perfect' wave is formed at the end of the domain. The closing wall is an outflow boundary, at which the analytical velocities are prescribed. This works very nicely in the case of wave propagation simulations without an object that disturbs the wave.

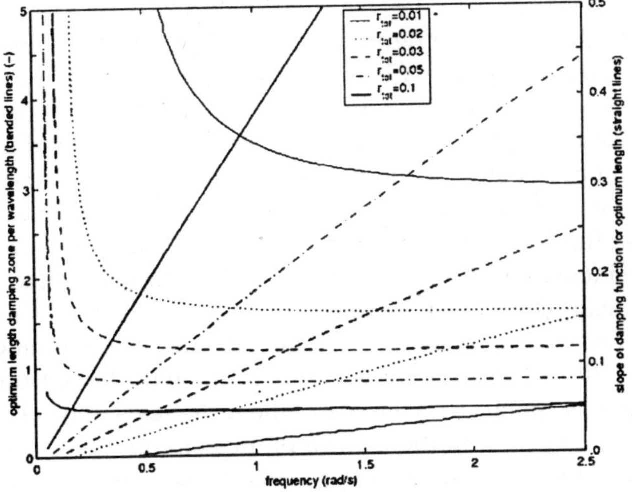

Fig. Combination of Required Length and Slope of the Dissipation Zone for a Certain Allowed Amount of Reflection.

- A combination of the dissipation zone with a Sommerfeld boundary condition is used. In this case, the closing wall is an outflow boundary where a Sommerfeld boundary condition has been applied. The Sommerfeld boundary condition is tuned for the smallest frequencies of the wave in the domain, whereas the dissipation function is tuned for the larger frequencies. In the current model we aim for the most general solution, so damping towards zero (the first option) is used. The wave is only damped in the travelling direction, whereas the tangential sides of the domain are solid walls, which have been placed at such a distance that the simulation results are not influenced by them.

When applying this to a rather high wave, some problems arise. The water level in the domain increases linearly in time, which is due to the fact that the net amount of water flowing into the domain at the inflow boundary is positive over one period and not zero, the so-called Stokes drift where the results of a wave simulation are shown with a period of 14.44 s, a wave height of 32.6 m, a wavelength of 325 m and the water depth is 600 m.

In the left of the figure the amount of water flowing into the domain is shown. Clearly, the integral over one period is not zero but positive. Because there is a solid wall at the end of the domain, the total amount of water is determined by the in- and outlet at the inflow boundary only, which causes the water level to increase linearly in time as is shown in the right of the figure. Because the wave is quite high, there is a very large increase of the water level, which is directly reflected in the mean of the wave elevation .

There are also some problems near the inflow boundary, where the wave is disturbed. Due to the reflections from the outflow boundary and the rise of the water level, the velocities and water height in the interior of the computational domain do not fit to the analytically prescribed velocities at the inflow boundary any more.

There are several ways to solve this increase of water problem. One way is to determine the net inflow and add an outflow boundary where this amount of fluid is forced out of the domain. A disadvantage of this method is, that it can only be applied to regular waves in a nice way, in which case it is known how much fluid should flow out of the domain at what time, due to the reguiarity.

A more flexible method has been found by changing the solid wall at the end of the domain into an outflow boundary, where hydrostatic pressure is prescribed. In a perfect situation, the water height at the end of the domain always equals the calm water level because of the damping of the wave. The fluid simulation will always try to maintain that level when the hydrostatic pressure for a water level equal to the water depth is prescribed at the outflow boundary.

Instead of the linear increase of water, a fluctuation of the water volume around the initial amount of water can be observed. The fluctuation is due to some reflections, which can be concluded from the fact that the period of the fluctuation is 8 wave periods, which means that the reflected wave travels 8 wavelengths before it reaches the starting position, the outflow boundary, again. In a domain consisting of 4 wavelengths, this is exactly 8 wavelengths (back and forth in the domain). The increase and decrease of the total water level during 8 periods is less than 0.5%.

FREE SURFACE VELOCITIES

The treatment of the velocities in the neighbourhood of the free surface is explained. It was concluded that special care has to be taken for the determination of SE-velocities, which are velocities at the cell face between a surface cell and an empty cell. Two methods are described, of which a combination is used in ComFLOW. In the first method SE-velocities are defined by demanding conservation of mass in an S-cell. This means that the total flux through the cell faces of the S-cell should be zero.

Another disadvantage of this method is the inaccuracy in wave simulations. The inaccurate prediction of an SE-velocity in case of a wave simulation can be understood from the right of the figure the horizontal velocity has been shown as function of the vertical coordinate. The theoretical values of the horizontal velocity in the neighbourhood of the free surface are indicated by the solid line.

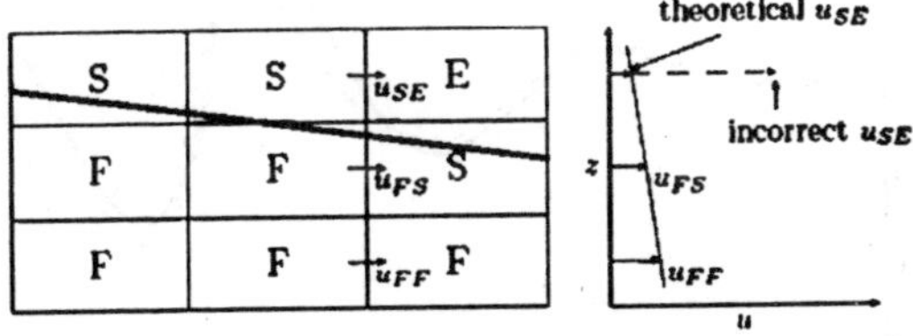

Fig. Inaccurate Prediction of USE in a Wave Simulation

To satisfy div(u) = 0 in the central S-cell, the SE-velocity uSE is copied from the left neighbour velocity (and the vertical velocity of this S-cell vSE is copied from the lower cell face).

Due to the coarseness of the grid, this neighbour velocity is about one or more meters left of the SE-velocity resulting in an inaccurate prediction of u_{SE} sketched by the dashed arrow.

In the second method for SE-velocities it is proposed to determine the velocities by choosing a direction, from which all SE-velocities in an S-cell are extrapolated. This direction is chosen as the direction where most fluid is present, so the coordinate direction that is 'most normal' to the free surface. In the case of waves, this is mostly the negative z-direction. As stated before, constant and linear extrapolation can be used. In wave simulations linear extrapolation gives superior results, since then the velocities are estimated very accurately. Here, an irregular steep wave propagates through an empty domain. A snapshot of the wave elevation is shown at the time point that the wave is most steep.

The asterisks show measurements of this wave. The mass conservation method does not predict the wave elevation accurately for this steep wave. Both constant and linear extrapolation produce a much better resemblance with the measurements. The simulation using linear extrapolation is most accurate. Based on the simulation of waves.

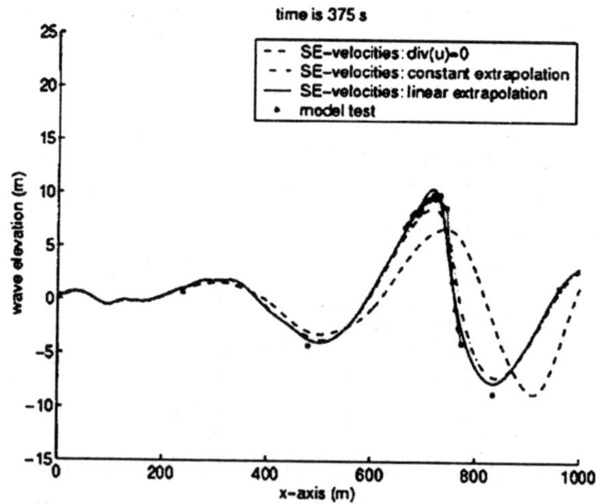

Fig. Wave Elevation of an Irregular Wave where Different SE-Velocity Treatments have been Compared: Mass Conservation in S-cells, Constant Extrapolation and Linear Extrapolation

WAVE PROPAGATION USING DIFFERENT OF METHODS

A few different VOF methods have been described and their performance has been tested using standard kinematic tests and a dambreak simulation. Four different methods have been used. First, the original Hirt-Nichols method has been used, where the interface is implicitly reconstructed in a piecewise constant manner.

To avoid flotsam and jetsam and to take care of mass conservation, a local height function has been introduced. A more sophisticated method for the displacement of the free surface is the method of Youngs. The free surface is reconstructed using piecewise linear elements and the displacement is based on this reconstruction. Youngs' method has also been used in combination with the local height function to ensure perfect mass conservation. The extrapolation method for free surface velocities has been used.

First, the different VOF methods are tested on the propagation of a regular wave. The wave has a period of 14.44 seconds and a length of 325 meter. The wave height is 10.14 meter and the water has a depth of 600 meter. At the outflow boundary a Sommerfeld condition is used. The resulting free surface profile after a simulation time of four wave periods. Also the error in the calculated free surface profile compared to the theoretical solution is shown, calculated by

$$E_w(x) = |\eta(x) - \eta th(x)| / H,$$

where η is the calculated wave elevation, ηth the theoretical wave elevation, and H the wave height. From the resulting error Ew, plotted in the bottom concluded that Youngs' method gives more accurate results in a regular wave simulation than Hirt-Nichols' method.

Youngs without a local height function is best in the interior of the domain, but has a larger error at the outflow boundary than Youngs with a local height function. No reason has been found for this behaviour. In all four simulations relatively little mass was lost. When examining the total amount of water in an area of 30 m high around the calm water level, all four methods have mass loss within 0.5% of the amount of water in that area. The calculation times of the four simulations are comparable. The factor between the calculation times of Youngs and Hirt-Nichols is 1.2.

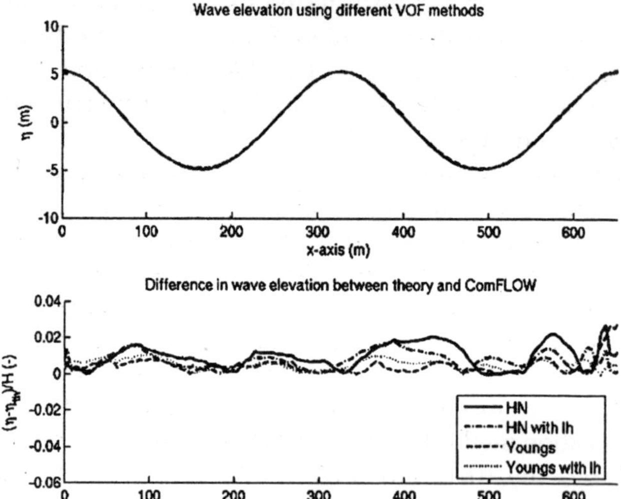

Fig. Wave Elevation (up) and Difference in Wave Elevation between Stokes 5th Order Theory and Simulation (down) for Different VOF Methods: Hirt-Nichols and Youngs Without and with a Local Height Function.

Youngs method does not take much extra calculation time in this simulation, since the interface reconstruction is only performed in a small percentage of the total number of cells (in average in 130 of the 6000 cells a reconstruction is made). The four different VOF methods are also used in the propagation of a steep wave event. The wave event is taken from an experiment at MARIN, to be more precise experiment 114002.

The wave is generated at the inflow boundary by prescribing a superposition of linear wave components derived from Fourier analysis of the measured wave elevation in the experiment. The wave elevation at the position and time point where the wave is high and steep. Youngs' method without a local height function results in the highest wave and gives the best agreement with the measurement.

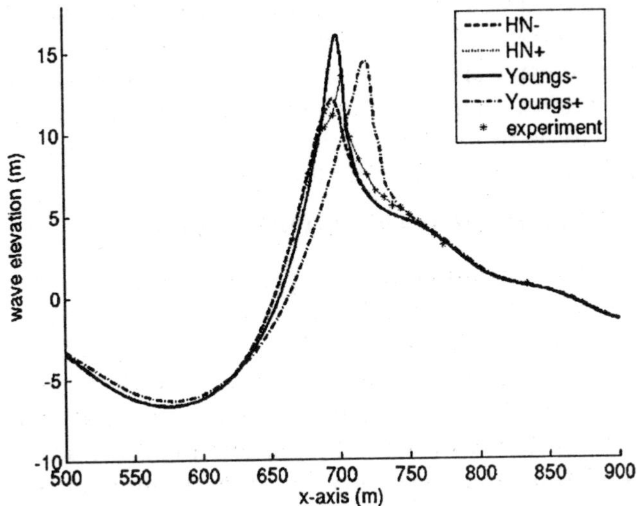

Fig. Wave Elevation of a Steep Wave Event using the Four Different VOF Methods:

Hirt-Nichols' VOF and Youngs' VOF with and without a local height function Hirt-Nichols with and without local height function give similar results, both are pretty good, but the wave crest is a bit flattened compared to the experiment. Youngs with a local height function performs worst, which is remarkable considering the good performance.

To conclude, Youngs' method performs best in wave simulations, its results are most accurate. A problem arises when combining Youngs' method with a local height function in the steep wave event. So, this combination should only be used with care until this problem is solved. Unfortunately, due to the choice of free surface velocities that are extrapolated, Youngs' method without local height function can cause loss of mass in the dambreak.

But in a simulation with a very smooth free surface as in the wave simulations, only a very small amount of water is lost using Youngs' method. So, for wave simulations, Youngs' method without local height function can be used without problem. But when performing a more violent simulation with a much distorted free surface, mass can be lost just mentioned. Mass can be perfectly conserved using Youngs' method, when it is combined with div(u) = 0 in surface cells. But using div(u) = 0 for the determination of free surface velocities results in inaccurate wave simulations and can cause instabilities in the computations. Therefore, this option should not be used. The safest way is to use Hirt-Nichols' method with local height function, but this can be a bit less accurate.

VALIDATION OF WAVE PROPAGATION

For the validation of wave propagation in several tests have been performed. Firstly, two-dimensional wave propagation without an object in the flow has been investigated. Attention has been paid to reflections at the outflow boundary, the influence of the artificial viscosity present due to the upwind discretisation and the size of the grid and time step necessary for an accurate simulation of waves.

Especially, steep and high waves, which are the most important waves in green water and wave impact calculations, have been studied. Secondly, simulations have been performed of wave loading on a spar platform. The waves are regular, very long and quite low. The simulation results have been compared with experimental results that have been provided by the Maritime Research Institute Netherlands (MARIN).

Two-Dimensional Wave Propagation: Regular Waves

A very extensive study of two-dimensional wave propagation without an object in the flow has been performed by Meskers. Some of the most important results will be shown here also. Attention will be paid to the number of cells and the time step that are needed for an accurate description of the wave. The simulations have been performed and compared with both the linear and the 5th order Stokes theory. Also, the influence of the artificial viscosity has been studied by performing a simulation of many wavelengths during many periods. These waves have been studied by simulating a design wave, which is a kind of mean shape of a wave in a linear random sea-state of which the power spectrum is given. Also a comparison of the use of a Sommerfeld boundary condition with a damping zone is made. The simulations have been performed for one example of a wave, which is one of the characteristic waves in an FPSO field with a water depth of 600 meter. This wave has a period of 14.44 seconds, resulting in a wave length of 325 meter. The wave height is varied between 10.14 and 20.28 meter. From the conclusions of Meskers in the characteristic parameter settings that have shown to give an accurate simulation of this example wave can also be used for deep water waves with different periods and wavelengths.

In the simulations of the wave shown in this thesis, a few numerical parameters have been varied, namely the number of time steps per period, the number of cells per wavelength, and the number of cells in the wave height. The results have been presented as the resulting wave elevation after a simulation time of four periods. The domain consists of four wavelengths in case of using a damping zone, of which two wavelengths are used as damping zone. The domain consists of two wavelengths when the Sommerfeld boundary condition is used.

Airy wave theory and 5th order Stokes theory have been used for the initial condition and the inflow boundary condition. In the lower picture the difference between the simulation and linear theory and Stokes 5th order theory, respectively, has been shown. Clearly, the 5th order Stokes results are much better than the linear Airy wave results, which is consistent with the fact that H/_ equals 0.06, outside the validity region of linear theory. First, the Sommerfeld outflow boundary condition is used that is based on the wave equation. Second, a damping zone is added as in the other simulations. Both methods give similar results.

The Sommerfeld boundary condition lets the wave flow out of the domain properly without much disturbance in the domain. The advantage of using the Sommerfeld condition over a damping zone is the number of grid cells that need to be used. In this simulation, the number of grid cells in the simulation using a damping zone is twice as large as the number of grid cells in the Sommerfeld simulation. On the other hand, the Sommerfeld condition can only be used in case of regular waves that are not too much disturbed. Further, the wave velocity should be given a priori, which is only possible when the wave characteristics are known.

The number of time steps per period has been altered. The difference between 250 and 500 time steps per period is not very large. From this it is concluded that 250 time steps per period is enough for an accurate simulation result.

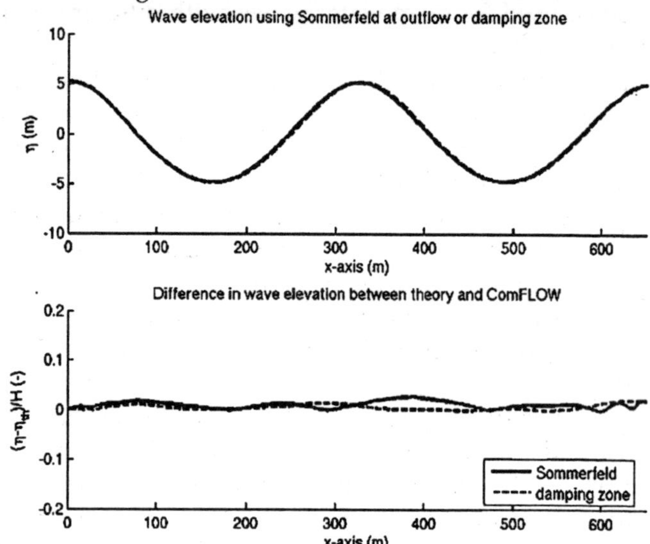

Fig. Wave Elevation (Top) and Difference in Wave Elevation between Theory and Simulation (Bottom) with Initialisation using Linear or Stokes 5th Order Theory

The accuracy of the wave simulation increases with a larger number of cells per wavelength. There is no difference between 60 and 90 cells per wavelength, so 60 cells per wavelength is enough to capture the wave of altering the number of cells in the wave height. The grid was equally stretched towards the calm water surface in the three different simulations.

There is not a large difference between the results. For all three simulations, the difference between computation and theory is about 5% of the wave height. In the case of 9 cells in the wave height, the simulation shows some small peaks, especially in the crest and trough of the wave. This is due to the ratio between dx and dz, which are the distances between two grid lines in x and z direction, respectively. Meskers gives in his report an estimate for the minimum required aspect ratio to prevent these wiggles, based on a series of simulations. In the future, this point should be investigated further to understand the nature of the wiggles. Dissipation of energy in wave simulations

Compared to a central discretisation the upwind discretisation can be interpreted as a central discretisation plus an extra diffusive term. This term adds extra viscosity to the physical viscosity.

The amount of extra viscosity is equal to uh/2 where u is the velocity and h the mesh size. So this term is dependent on the position in space. When simulating waves, we have to investigate the influence of the artificial diffusion on the wave propagation. It can be seen that the influence is definitely not very large over a few periods, because no real damping is visible after the four periods that have been simulated. To get a better idea of the influence of the artificial viscosity, a wave has been simulated for many periods in a domain of 25 wavelengths.

The wave has a period of 1.9 seconds and a wave height of 0.16 meter. The wave elevation as function of time and as function of distance to the inflow boundary are shown. The wave elevation as function of time at a position 14 wavelengths from the inflow boundary is shown. The simulation is started with an undisturbed (and thus undamped) wave field. During the simulation the amplitude of the wave decreases, until the stationary damping is reached after about 58 seconds. At this position, 14 wavelengths from the inflow boundary, the wave height has decreased from 0.16 meter to 0.125 meter, which is a decrease of 22% of the wave height.

From the right picture it can be seen that the wave height has decreased 28% at 20 wavelengths behind the inflow boundary after a simulation time of 100 periods. An estimate of the maximum artificial viscosity that is added due to the upwind discretisation can be found by calculating uh/2. For the velocity u the maximum of the mean velocity over one period is taken, which according to linear theory is equal to $u = gAk/\omega \approx 0.26$m/s. So the approximate maximum artificial viscosity is given by $k_{art} = uh/2 \approx 0.26 * 140/1500/2 = 0.012$ n.

Where as the physical kinematic viscosity v is equal to 10^{-6} m^2/s is really caused by the artificial viscosity, simulations have been performed with less artificial viscosity (only 10% of the artificial viscosity is added) and with a central discretisation. The simulation with the central discretisation has been performed for 58 periods in a domain of 12 wavelengths, which is because the grid sizes and time step have to be very small in a calculation with a central discretisation.

The wave elevation at 9 wavelengths behind the inflow boundary has been given for the different amounts of added viscosity. It can be clearly seen, that the amplitude of the wave is decreasing more in time when the added viscosity becomes larger where the wave height decrease is shown after 100 periods of simulation at the different locations behind the inflow boundary. We can conclude from this figure that there is a large influence of the artificial viscosity, which has a damping effect on the wave.

But clearly, this is not the only reason for dissipation of energy in the wave simulations. Although in the central discretisation no artificial viscosity is present, the wave has been damped much more than it should be according to the physics. This is due to the boundary conditions and the displacement of the free surface that sometimes are chosen to be a bit dissipative to get a stable solution. Summarising, for wave simulation in a very long domain, where many periods are simulated, a clear damping is visible.

This dissipation of energy is for a great deal due to the artificial viscosity that is added when using an upwind discretisation. But another part of the energy dissipation is coming from the treatment of the boundaries and the free surface. Although an energy preserving discretisation is used , some energy is lost in other parts of the algorithm. The loss of energy is only a few percent in a small domain when not that many periods are simulated. So for the applications of wave loading on ships, the influence will not be very significant.

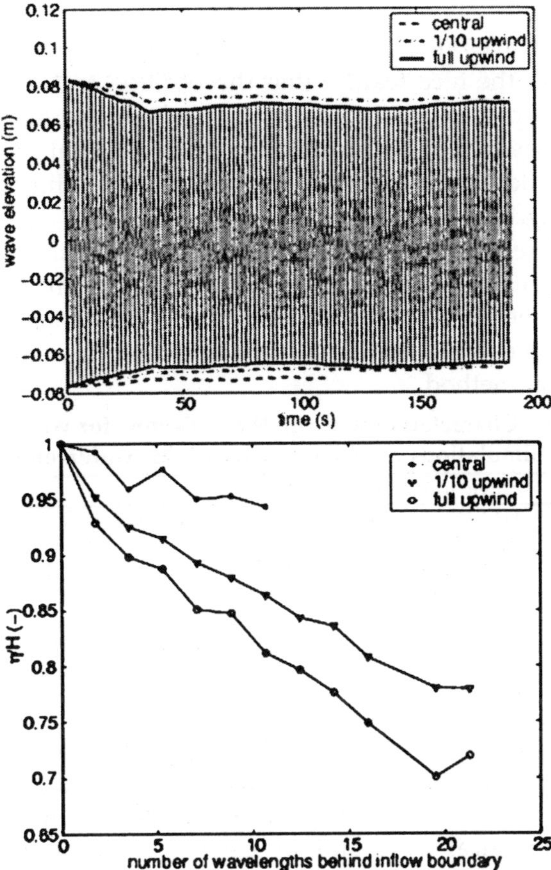

Fig. Left: Wave Elevation at 10 Wavelengths Behind the Inflow Boundary; Right: Decrease of Wave Height after A Simulation Time of 100 Periods at Different Distances from the Inflow Boundary

Two-Dimensional Wave Propagation: Steep Wave Events

For the validation of irregular steep or high waves, waves generated on basis of Newwave theory have been used such a wave event has been developed as a mean wave shape of a high wave in a given linear random sea-state. The wave is modelled by a superposition of linear waves, and thus is a linear wave. High waves are not linear of form, and therefore, mostly some nonlinear corrections are being made to the waves.

The waves are sometimes used as a design wave, because they have the nice feature that the position and time of the largest impact can be predicted beforehand. At marin these design waves have been used in the experimental program of the Safe flow project to get a better understanding of wave slamming.

Before the experiments with a vessel in the waves is performed, the waves have been calibrated in the basin without the presence of a vessel. The measurements of this undisturbed wave elevation have been used to compare our numerical method.

Table. Characteristics of the Wave Events, for which the Simulations are Compared with Measurements

Test no.	*Single wave event description*
112003	1/14 stepp wave in sea state steepness for 100 year return period
113001	mean highest wave for 100 year return period
114002	1/16 steep wave in sea state steepness for 100 year return period
119001	1/18 steep wave in sea state steepness for 100 year return period
120001	highest wave in design wave spectrum with Hm_0 = 12m, T_0 = 12*s*, γ = 2.5.
121001	steepest wave in design wave spectrum with Hm_0 = 12*m*, T_0 = 12*s*, γ = 2.5.

The waves have been generated at the inflow boundary of the numerical wave tank by using a Fourier transform of the time trace of a measured wave height. The calculated wave elevation in the domain is compared to the measurements at positions of 240, 480, 720, and 960 m behind the inflow boundary.

Detailed measurements, using 19 wave probes, have been performed at the position where the wave is expected to be highest between 666 and 774 m behind the inflow boundary. The calculated wave height of test 112003 at 720 m behind the inflow boundary, where the wave is expected to be highest, is compared to theory and experiment. The comparison with the experiment is very good. The amplitude of the simulation is a bit smaller than in the experiment, which has been observed before as a consequence of using Hirt-Nichols' method for the free surface displacement.

Also the dissipation of energy due to the upwind discretisation could have an influence on the wave height. The simulation has been performed on two different grids, using 600 × 50 grid cells and 1200 × 100 grid cells. The grid refinement does not have a large influence, almost no difference can be observed between the two grids. The wave elevation predicted from linear theory, extrapolated from the FFT of the measured wave elevation, is shown, which is very different from the calculations and measurement. This means that ComFLOW has correctly dealt with the nonlinearities that are present in reality.

The simulations of the other events have all been performed with the fine grid of 1200x100 grid points are much the same as the results of the first test. The wave does not completely reach the height of the experiment in the first peak. This is confirmed by the wave profile at 430 s, where the wave has collapsed a little. Again, the simulations do not reproduce the exact height of the first peak at about 395 seconds in both tests. Nevertheless, the other parts of the simulation are fairly close to the experiment.

To be sure that the wave is well reproduced in the simulation the wave must be compared with the undisturbed wave in the basin first. When it has been confirmed that the wave is the same as in the basin, it can be used for a simulation of wave impact. Further, the results show that ComFLOW reproduces nonlinearities from the experiment, but does not always reproduce the height of the wave exactly.

Regular Wave Loading on a Spar Platform

For the validation of wave loading, simulations have been performed of regular wave loading on a spar platform. A spar platform is a floating structure for drilling and the production of crude oil in the ocean. Typically, a spar is a long cylindrical steel structure with 30-50 meters in diameter and 200 meters in length.

At MARIN experiments with a spar have been performed, where the spar was fixed while regular waves hit the structure. In full scale the spar has a total length of 220 m and a diameter of 35 m, the draft is 200 m in a water depth of 290.35 m. The spar has been divided into three horizontal segments, on which forces have been measured.

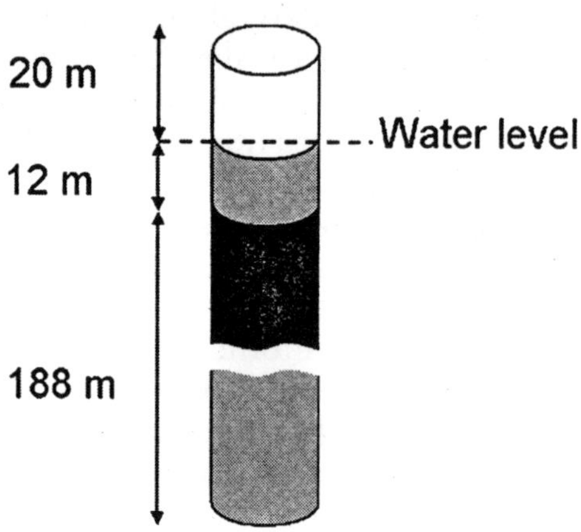

Fig. Configuration of the Spar Buoy, which is Divided into 3 S●segments: the Part above the Calm Water Surface, the Part 12 Meter Below the Calm Water Surface and the Remainder

The spar buoy has been placed at a distance of one wavelength behind the inflow boundary.

Table. Wave Characteristics of Simulations of Loading on a Spar Platform

Test No.	*Frequency*	*Period*	*Wave height*	*Wavelength*
101	0.26 rad/s	24.17s	10.77 m	866 m
201	0.26 rad/s	24.17s	23.62 m	866 m
001	0.48 rad/s	13.09s	10.91 m	274 m

About 100 m behind the cylinder, a dissipation zone of one wavelength has been added to prevent the wave from reflecting into the domain.

Chapter 7

Lasers

"Laser" is an acronym for "Light Amplification by Stimulated Emission of Radiation", coined in 1957 by the laser pioneer Gordon Gould. Although this original meaning denotes an principle of operation, the term is now mostly used for *devices* generating light based on the laser principle.

The first laser device was a pulsed ruby laser, demonstrated by Theodore Maiman in 1960. In the same year, the first gas laser (a helium–neon laser) and the first laser diode were made.

The term "optical maser" (MASER = microwave amplification by stimulated amplification of radiation) was initially used, but later replaced with "laser".

Laser technology is at the core of the wider area of photonics, essentially because laser light has a number of very special properties:

- It is usually emitted as a laser beam which can propagate over long lengths without much divergence and can be focused to very small spots.
- It can have a very narrow bandwidth, whereas e.g. most lamps emit light with a very broad spectrum.
- It may be emitted continuously, or alternatively in the form of short or ultrashort pulses, with durations from microseconds down to a few femtoseconds.

These properties, which make laser light very interesting for a range of applications, are to a large extent the consequences of the very high degree of coherence of laser radiation.

LASER WORKS

A laser usually comprises an optical resonator (laser resonator, laser cavity) in which light can circulate (e.g. between two mirrors), and within this resonator a gain medium (e.g. a laser crystal), which serves to amplify the light. Without the gain medium, the circulating light would become weaker and weaker in each resonator round trip, because it experiences some losses, e.g. upon reflection at mirrors.

However, the gain medium can amplify the circulating light, thus compensating the losses if the gain is high enough. The gain medium requires some external supply of energy – it needs to be "pumped", e.g. by injecting light (*optical pumping*) or an electric current (*electrical pumping* '! *semiconductor lasers*). The principle of laser amplification is stimulated emission.

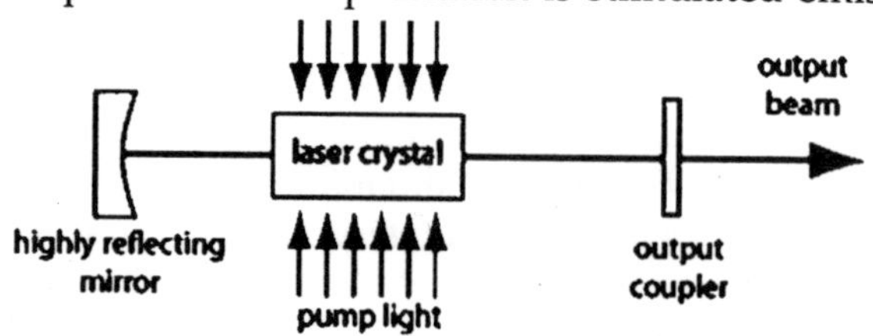

Fig Setup of a Simple Optically Pumped Laser.

The laser resonator is made of a highly reflecting curved mirror and a partially transmissive flat mirror, the output coupler, which extracts some of the circulating laser light as the useful output. The gain medium is a laser crystal, which is side-pumped, e.g. with light from a flash lamp. A laser can not operate if the gain is smaller than the resonator losses; the device is then below the so-called laser threshold and only emits some luminescence light. Significant power output is achieved only for pump powers above the laser threshold, where the gain can exceed the resonator losses.

If the gain is larger than the losses, the power of the light in the laser resonator quickly rises, starting e.g. with low levels of light from fluorescence. As high laser powers saturate the gain, the laser power will in the steady state reach a level so that the saturated gain just equals the resonator losses (' *gain clamping*). Before reaching this steady state, a laser usually undergoes some relaxation oscillations. The threshold pump power is the pump power where the small-signal gain is just sufficient for lasing.

Some fraction of the light power circulating in the resonator is usually transmitted by a partially transparent mirror, the so-called output coupler mirror. The resulting beam constitutes the useful output of the laser. The transmission of the output coupler mirror can be optimized for maximum output power.

Some lasers are operated in a continuous fashion, whereas others generate pulses, which can be particularly intense. There are various methods for pulse generation with lasers, allowing the generation of pulses with durations of microseconds, nanoseconds, picoseconds, or even down a few femtoseconds (↔ *ultrashort pulses* from *mode-locked lasers*).

The optical bandwidth (or linewidth) of a continuously operating laser may be very small when only a single resonator mode can oscillate ('! *single-frequency operation*). In other cases, particularly for mode-locked lasers, the bandwidth can be very large – in extreme cases, it can span about a full octave. The center frequency of the laser radiation is typically near the frequency of maximum gain, but if the resonator losses are made frequency-dependent, the laser wavelength can be tuned within the range where sufficient gain is available. Some broadband gain media such as Ti:sapphire and Cr:ZnSe allow wavelength tuning over hundreds of nanometers.

Due to various influences, the output of lasers always contains some noise in properties such as the output power or optical phase.

Types of Lasers

Common types of lasers are:

- *Semiconductor lasers* (mostly laser diodes), electrically (or sometimes optically) pumped, efficiently generating very high output powers (but typically with poor beam quality), or low powers with good spatial properties (e.g. for application in CD and DVD players), or pulses (e.g. for telecom applications) with very high pulse repetition rates. Special types include quantum cascade lasers (for mid-infrared light) and surface-emitting semiconductor lasers (VCSELs and VECSELs), the latter also being suitable for pulse generation with high powers.

- *Solid-state lasers* based on ion-doped crystals or glasses (*doped insulator lasers*), pumped with discharge lamps or laser diodes, generating high output powers, or lower powers with very high beam quality, spectral purity and/or stability (e.g. for measurement purposes), or ultrashort pulses with picosecond or femtosecond durations. Common gain media are Nd:YAG, Nd:YVO_4, Nd:YLF, Nd:glass, Yb:YAG, Yb:glass, Ti:sapphire, Cr:YAG and Cr:LiSAF. A special type of ion-doped glass lasers are:
- *Fiber lasers*, based on optical glass fibers which are doped with some laser-active ions in the fiber core. Fiber lasers can achieve extremely high output powers (up to kilowatts) with high beam quality, allow for widely wavelength-tunable operation, narrow linewidth operation, etc.
- *Gas lasers* (e.g. helium–neon lasers, CO_2 lasers, and argon ion lasers) and excimer lasers, based on gases which are typically excited with electrical discharges. Frequently used gases include CO_2, argon, krypton, and gas mixtures such as helium–neon. Common excimers are ArF, KrF, XeF, and F_2.

Less common are chemical and nuclear pumped lasers, free electron lasers, and X-ray lasers.

Laser Sources in a Wider Sense

There are some light sources which are not strictly lasers, but are nevertheless often called *laser sources*:

- In some cases, the term is used for amplifying devices emitting light without an input (excluding seeded amplifiers). An example is X-ray lasers, which are usually *superluminescent sources,* based on spontaneous emission followed by single-pass amplification. There is then no laser resonator. A similar situation occurs for optical parametric generators, where the amplification, however, is not based on stimulated emission. Light from such devices can have laser-like properties, such as strongly directional emission and a limited optical bandwidth.

- In other cases, the term *laser sources* is justified by the fact that the source contains a laser, among other components.
  This is the case for combinations of lasers and amplifiers ('! *master oscillator power amplifier*), and also for sources based on nonlinear frequency conversion of laser radiation, e.g. with frequency doublers or optical parametric oscillators.

SAFETY ASPECTS

The work with lasers can raise significant safety issues. Some of those are directly related to the laser light, in particular to the high optical intensities achievable, but there are also other hazards related to laser y for details.

Laser Light

Laser light (laser radiation) is simply light generated with a laser device. Such light has some very special properties, which very much distinguish it from light with other origins:

- Laser light is usually delivered in the form of a laser beam, i.e. it propagates dominantly in a well-defined direction with moderate beam divergence. Such a laser beam has a high (sometimes extremely high) degree of spatial coherence.
  This means that the electric fields at different locations across a beam profile oscillate with a rigid phase relationship. Exactly this coherence is the reason why a laser beam can propagate over long distances without spreading very much in the

transverse directions, and why it can be focused to very small spots (high *focusability* of laser beams).

- In many but not all cases, laser light also has a high degree of *temporal* coherence, which is equivalent to a long coherence length. This means that a rigid phase relationship is also maintained over relatively long time intervals, corresponding to large propagation distances (often many kilometers) or to huge numbers of oscillation cycles.
- The large temporal coherence, quantified with a large coherence time or coherence length, is associated with a narrow spectral bandwidth (or linewidth). For a visible laser beam, this means that it has a certain pure color, e.g. red, green or blue, but not white or magenta. Some lasers allow a degree of wavelength tuning (e.g. dye lasers). The large coherence length introduces a tendency for the phenomenon of laser speckle, i.e. a characteristic granular pattern which can be observed e.g. when the laser beam hits a metallic surface.
- In most cases, laser light is linearly polarized. This means that the electric field oscillates in a particular spatial direction (→ *polarization of laser emission*).

Depending on the case, laser light can have other remarkable properties:

- Laser light may be visible, but most lasers actually emit in other spectral regions, in particular in the near-infrared region, which human eyes cannot perceive.
- Laser light is not always continuous, but may be delivered in the form of short or ultrashort pulses. As a consequence, the peak power can be extremely high – for some amplified laser systems well above 1 TW (10^{12} W) – if the pulse duration is far below the temporal distance of the pulses, so that a large pulse energy is possible.
- The noise properties of lasers can also be very interesting (→*laser noise*). For example, the oscillation

frequency of a laser can be stabilized to stay within an extremely narrow range.

For comparison, light from an incandescent lamp is partly visible and mostly infrared, has a very high spectral bandwidth, cannot be strongly focused (because of low spatial coherence), and cannot be generated in the form of short pulses.

Laser light can cause significant hazards, particularly for the eye, but also for the skin, apart from fire hazards and others. Lasers pulses are often particularly dangerous, because they can have enormously high optical intensities, and invisible laser beams add specific risks..

MANUFACTURING

Lasers are widely used in manufacturing, e.g. for cutting, drilling, welding, cladding, soldering, hardening, ablating, surface treatment, marking, engraving, micromachining, pulsed laser deposition, lithography, alignment, etc. In most cases, relatively high optical intensities are applied to a small spot, leading to intense heating, possibly evaporation and plasma generation. Essential aspects are the high spatial coherence of laser light, allowing for strong focusing, and often also the potential for generating intense pulses.

Laser processing methods have many advantages, compared with mechanical approaches. They allow the fabrication of very fine structures with high quality, avoiding mechanical stress such as caused by mechanical drills and blades. A laser beam with high beam quality can be used to drill very fine and deep holes, e.g. for injection nozzles. A high processing speed is often achieved, e.g. in the fabrication of filter sieves. Further, the lifetime limitation of mechanical tools is removed. It can also be advantageous to process materials without touching them.

Medical Applications

There is a wide range of medical applications. Often these relate to the outer parts of the human body, which are easily reached with light; examples are eye surgery and vision correction (LASIK), dentistry, dermatology (e.g. photodynamic therapy of cancer), and various kinds of cosmetic treatment such as tattoo removal and hair removal.

Lasers are also used for surgery (e.g. of the prostate), exploiting the possibility to cut tissues while causing minimal bleeding.

Very different types of lasers are required for medical applications, depending on the optical wavelength, output power, pulse format, etc. In many cases, the laser wavelength is chosen such that certain substances (e.g. pigments in tattoos or caries in teeth) absorb light more strongly than surrounding tissue, so that they can be more precisely targeted.

Medical lasers are not always used for therapy. Some of them rather assist the diagnosis, e.g. via methods of laser microscopy or spectroscopy.

Metrology

Lasers are widely used in optical metrology, e.g. for extremely precise position measurements with interferometers, for long-distance range finding and navigation. Laser scanners are based on collimated laser beams, which can read e.g. bar codes or other graphics over some distance. It is also possible to scan three-dimensional objects, e.g. in the context of crime scene investigation (CSI).

Optical sampling is a technique applied for the characterization of fast electronic microcircuits, microwave photonics, terahertz science, etc.

Lasers also allow for extremely precise time measurements and are therefore essential component of optical clocks which are beginning to outperform the currently used cesium atomic clocks. Fiber-optic sensors, often probed with laser light, allow for the distributed measurement of temperature, stress, and other quantities e.g. in oil pipelines and wings of airplanes.

Data Storage

Optical data storage e.g. in compact disks (CDs), DVDs, Blu-ray Discs and magneto-optical disks, nearly always relies on a laser source, which has a high spatial coherence and can thus be used to address very tiny spots in the recording medium, allowing a very high density data storage. Another case is holography, where the temporal coherence can also be important.

Communications

Optical fiber communication, extensively used particularly for long-distance optical data transmission, mostly relies on laser light in optical glass fibers. Free-space optical communications, e.g. for inter-satellite communications, is based on higher-power lasers, generating collimated laser beams which propagate over large distances with small beam divergence.

Displays

Laser projection displays containing RGB sources can be used for cinemas, home videos, flight simulators, etc., and are often superior to other displays concerning possible screen dimensions, resolution and color saturation. However, further reductions in manufacturing costs will be essential for deep market penetration.

Spectroscopy

Laser spectroscopy is useful, e.g. in atmospheric physics and pollution monitoring (e.g. trace gas sensing with differential absorption LIDAR technology). It also plays a role in medicine (e.g. cancer detection), biology, and various types of fundamental research, partly related to metrology.

Microscopy

Laser microscopes and setups for coherence tomography provide images of, e.g., biological samples with very high resolution, often in three dimensions. It is also possible to realize functional imaging.

Various Scientific Applications

Laser cooling makes it possible to bring clouds of atoms

or ions to extremely low temperatures. This has applications in fundamental research and also for industrial purposes.

Particularly in biological and medical research, optical tweezers can be used for trapping and manipulating small particles, such as bacteria or parts of living cells.

Laser guide stars are used in astronomical observatories in combination with adaptive optics for atmospheric correction. They allow substantially increased image resolution even in cases where a sufficiently close-by natural guide star is not available.

Energy Technology

In the future, high-power laser systems might play a role in electricity generation. Laser-induced nuclear fusion is investigated as a alternative to other types of fusion reactors. High-power lasers can also be used for isotope separation.

Military Applications

There are a variety of military laser applications. In relatively few cases, lasers are used as weapons; the "laser sword" has become popular in movies, but not in practice. Some high-power lasers are currently developed for potential use as *directed energy weapons* on the battle field, or for destroying missiles, projectiles and mines.

In other cases, lasers function as target designators or laser sights (essentially laser pointers emitting visible or invisible laser beams), or as irritating or blinding (normally not directly destroying) countermeasures e.g. against heat-seeking anti-aircraft missiles. It is also possible to blind soldiers temporarily or permanently with laser beams, although the latter is forbidden by rules of war.

There are also many laser applications which are not specific for military use, e.g. in areas such as range finding, LIDAR, and optical communications.

Laser Design

The term *design* can have two different meanings. In some cases, it is meant to be a detailed description of a device, including e.g. used parts, how they are put together, and important operation parameters. In other cases, the term denotes the process leading to such a description.

Defining the Design Goals

Before a design is made, the *design goals* must be carefully evaluated. These should include not only the central performance parameters such as output power and wavelength; many more details can be relevant:

In Short:Details of the design of a laser product can have a strong impact on its performance, reliability, flexibility, and manufacturing cost. It is essential to be aware of all the design goals, to know the design properties required for reaching them, and to use a proper laser design (which is worked out at the beginning) as a vital part of an efficient development process.

- Optimum performance, e.g. in terms of output power, power efficiency, beam quality, brightness, intensity and/or phase noise, long-term stability (e.g. of the output power or the optical frequency), timing jitter, etc.
- Compact and convenient setup, ease of operation (e.g. simple turn-on procedure, simple wavelength tuning, no need for realignment)
- Maximum flexibility (e.g. for changing operation parameters)
- Reliability, low maintenance requirements, simple and cost-effective error analysis, maintenance and repair
- Minimum sensitivity to vibrations, temperature changes, electromagnetic interference, aging of components
- Low production cost, i.e., a small number of parts, simple alignment and testing, avoiding the use of parts which are expensive, sensitive, or difficult to obtain

It is certainly advisable to work out carefully the list of these requirements for the particular case before investing any significant resources in laser development, because it can easily be much more expensive and time-consuming to introduce additional properties into an already existing device.

IMPORTANT ASPECTS OF LASER DESIGNS

The properties of the designed laser device are largely determined by the design details, not only by the parts used. Some aspects are particularly important:

- General design parameters, such as resonator length (influencing compactness, tuning issues, frequency stability, etc.), pump intensity
- Selection of gain medium (e.g. a laser crystal) and pump source, suitable choice of geometry (e.g. rod or thin disk, side pumping or end pumping), doping concentration, crystal length, etc.
- Pump setup (e.g. for diode-pumped lasers), influencing output power and beam quality, long-term stability, and the ease of exchanging pump diodes
- Optimum type of laser resonator (e.g. as linear or ring laser, with monolithic or with discrete elements) and optimized resonator design, influencing aspects such as the number of parts, the output power and beam quality, alignment tolerances, sensitivity to thermal lensing, mechanical stability and drifts
- Selection and placement of laser mirrors and intracavity components for wavelength tuning, generation of short pulses via mode locking, dispersion compensation, frequency stabilization, etc.
- Mechanical housing, influencing mechanical stability, efficiency of cooling, temperature drifts, ease of maintenance, and safety issues
- Electronic equipment, e.g. for stabilizing the output power, controlling the laser wavelength, monitoring the status of pump diodes or temperatures, ensuring safe operation
- Proper documentation, including a part list (possibly with suppliers), mechanical designs, alignment and

testing procedures, design ideas, possibly optional extensions and limitations for modifying operation parameters

This list, which is certainly not yet complete, shows that proper laser designs are not a trivial matter, but are essential for achieving full customer satisfaction, cost efficiency, and flexibility for future developments.

Needed for Designing Lasers

Designing a laser is a challenging task. The following are definitely required:

- A detailed understanding of the requirements, which may include some understanding of the application
- A detailed understanding of all the relevant physical effects, such as laser amplification, thermal lensing, resonator modes, laser noise, etc., and their interaction
- The essential data, e.g. of laser crystals
- The ability to reduce the complexity to a practical level without losing important details
- Flexible software for calculations and simulations
- Practical experience with lasers, enabling one to recognize typical problems, correctly interpret experimental observations, etc.

It is clear that software alone is by no means sufficient to work out good laser designs.

Role of a Design in a Development Project

It is common practice, but nevertheless generally not advisable, to consider a laser design as a result of a development process which is largely based on trial and error. The design then plays a minor role, just summarizing the results of a lengthy process.

In such cases, the design is often not even properly documented, which creates a risk of losing a lot of potentially valuable information while saving only a minor amount of time at the moment.

In any non-trivial design project – and laser design

projects are hardly ever trivial – it is very advisable to attribute a vital role to the laser design:

- The laser design *is made in the office,* not in the laboratory, and properly deals with all known issues which can be or become relevant. Although this process can be much faster and cheaper than a trial-and-error approach in the lab, it takes some considerable discipline and of course requires a comprehensive expertise.
- A proper design is not just a set of ideas, but a very specific description, including e.g. the list of required parts, a more or less detailed prescription on how to put them together wherever this is not trivial, and is ideally supplemented by a description of the underlying reasoning, a discussion of limitations, etc.
- The prototype is then fabricated according to the design, and not vice versa. This greatly speeds up the fabrication, thus making efficient use of costly laboratory resources.
- For any future development of similar kind, the carefully worked out design will be a very valuable input. If it does not exist, and particularly if in addition a vital person has left the company, future developments will be much less efficient.

Attempts to abbreviate this process carry the risk of obtaining reduced performance values and of large time delays due to unexpected technical problems. The later such problems are recognized, understood and solved, the larger can be the resulting damage.

Deriving Designs from Older Designs

In industrial development, it is common to derive some product design from an older design, rather than starting from scratch.

Although this appears very economical, there are significant risks, particularly in cases where the first design has not be properly worked out and documented in a process as described above. A central challenge is that modifying some detail of a laser design may easily have unexpected side effects, introducing new problems which then require additional measures, which again can have side effects.

For such reasons, starting with some initial design, which works e.g. with some lower than desired output power, can be helpful, but it still requires a detailed understanding of that design and its limitations. A proper design document for the initial design can make it easy to produce a whole family of designs, which differ in e.g. output power or pulse repetition rate.

LASER RESONATORS

A laser requires a *laser resonator* (or *laser cavity*), in which the laser radiation can circulate and pass a gain medium which compensates the optical losses. Exceptions are only exotic cases where a medium with very high gain is used, so that amplified spontaneous emission extracts significant power in a single pass through the gain medium.

The laser radiation is automatically generated at one or several frequencies corresponding to resonances (resonator modes), possibly with small deviations caused by "gain pulling". No special measures are required for operating on the resonance; this is different for external resonators, e.g. resonant enhancement cavities.

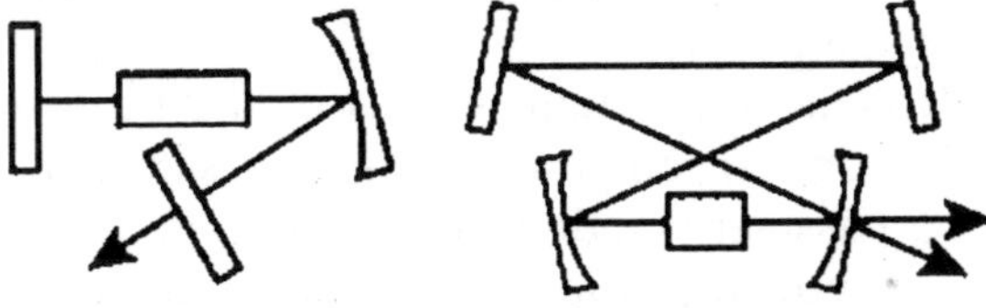

Fig. Two Simple Solid-State Laser resonators with a Laser Crystal as Gain Medium.

Output beams are generated where resonator mirrors are partially transmissive.

The ring laser (right) can exhibit laser action in two directions, thus generate two output beams; unidirectional operation could be enforced, e.g. with an additional intracavity optical isolator.

LASER RESONATORS OF SOLID-STATE LASERS

Solid-state bulk lasers are usually built with several dielectric mirrors (laser mirrors), which may be plain or curved. A linear resonator and a ring resonator built in that way, and containing a laser crystal as the gain medium. In some cases, a dielectric mirror coating is placed on the gain medium itself on monolithic solid-state lasers. One of the mirrors, usually an end mirror, is the partially transmissive output coupler.

The design of the laser resonator (comprising optical elements, angles of incidence, and distances between the components) determines the beam radius of the fundamental mode at all locations along the beam, together with other important properties. For maximum beam quality (→ *diffraction-limited* output), the beam radius in the gain medium has to match approximately the radius of the pumped region.

For smaller beam radii, operation with multiple spatial modes is obtained, leading to a non-ideal beam quality; however, such multimode lasers have other advantages such as much wider stability zones and a lower sensitivity to misalignment.

In many cases, the laser resonator design should have additional features. For example, it can be optimized

- For compactness
- For achieving certain values of the beam radius in other optical components (e.g. on a saturable absorber of a passively mode-locked laser)
- For avoiding too small beam radii in optical components (leading to optical damage particularly in Q-switched lasers)
- For minimizing adverse effects of thermal lensing and aberrations in the gain medium
- For minimizing the alignment sensitivity
- For accommodating multiple laser heads
- For a certain resonator length, determining the pulse repetition rate in a mode-locked laser or the pulse duration of a Q-switched laser.

Particularly for high-power lasers with good beam quality, thermal lensing in the gain medium is very important. The resonator design should be made so that changes of the thermal lens do not affect too much the mode sizes. Also, it should have a low sensitivity to thermal aberrations and misalignment . The importance of these factors should not be underestimated; there are cases where two resonators even with equal mode sizes in the gain medium lead to very different laser performance and are radically different in terms of alignment.

Although it is normally not that difficult to evaluate the properties of a given laser resonator, it can be challenging to find a resonator design which satisfies multiple criteria such as those listed above. Numerical optimization, using special resonator design software, can be the only way to find good solutions, particularly for some mode-locked lasers. Also, a solid understanding of resonator properties can help considerably when trying to find resonator configurations with special combinations of properties, such as large mode areas and short lengths. For advanced design issues, a great deal of experience is at least as important as a versatile design software.

Some high-power lasers (for example with slab designs) are operated with *unstable resonators*, allowing a reasonable (but typically not diffraction-limited) beam quality to be achieved despite the presence of strong thermal effects in the gain medium. Due to the high diffraction losses, such laser cavities require relatively high gain.

There are various types of monolithic solid-state lasers which have the whole beam path within the laser crystal. Beam reflections are then typically realized either with dielectric coatings on crystal surfaces, or with total internal reflection.

Laser Crystals

Laser crystals are typically single crystals (monocrystalline material) which are used as gain media for solid-state lasers. In most cases, they are doped with either trivalent rare earth ions or transition metal ions. These ions enable the crystal to amplify light at the laser wavelength via stimulated emission, when energy is supplied to the crystal via absorption of pump light (→ *optical pumping*). Compared with doped glasses, crystals usually have higher transition cross sections, a smaller absorption and emission bandwidth, a higher thermal conductivity, and possibly birefringence. In some cases, monocrystalline laser materials may be replaced with ceramic gain media, which have a fine polycrystalline structure.

Common Laser-active Dopants

The most frequently used laser-active rare earth ions and host media together with some typical emission wavelengths:

Table. Common rare Earth Ions in Laser–Active Crystals.

Ion	*Common host crystals*	*Important emission wavelengths*
neodymium 1053, (Nd^{3+})	$Y_3Al_5O_{12}$ (YAG), $YAlO_3$ (YALO), YVO_4(yttrium vanadate), $YLiF_4$ (YLF), tungstates ($KGd(WO_4)_2$, $KY(WO_4)_2$)	1064, 1047, 1342, 946 nm
ytterbium (Yb^{3+})	YAG, tungstates (e.g. KGW, KYW, KLuW), YVO_4, borates (BOYS, GdCOB), apatites (SYS), sesquioxides (Y_2O_3, Sc_2O_3)	1030, 1020–1070 nm
erbium (Er^{3+})	YAG, YLF	2.9, 1.6 μm
thulium (Tm^{3+})	YAG	1.9–2.1 μm
holmium (Ho^{3+})	YAG	2.1, 2.94 μm
cerium (Ce^{3+})	YLF, LiCAF, LiLuF, LiSAF, and similar fluorides	0.28–0.33 μm

The following tables lists common transition-metal doped crystals:

Table. Common Transition Metal ions in Laser–Active Crystals.

Ion	*Common host crystals*	*Important emission wavelengths*
titanium (Ti^{3+})	sapphire	650–1100 nm
chromium (II) (Cr^{2+})	zinc chalcogenides such as ZnS, ZnSe, and ZnS_xSe_{1-x}	2–3.4 μm
chromium (III) (Cr^{3+})	Al_2O_3 (ruby), $LiSrAlF_6$ (LiSAF), LiC aAlF$_6$ (LiCAF), $LiSrGaF_6$ (LiSGAF)	0.8–0.9 μm
chromium (IV)	YAG, $MgSiO_4$ (forsterite)	1.35–1.65 (Cr^{4+}) μm (YAG), 1.1–1.37 μm forsterite)

These tables contain only the most common host crystals; many others exist, which are less frequently used.

Important Properties of Host Crystals

The host crystal is much more than just a means to fix the laser-active ions at certain positions in space. A number of properties of the host material are important:

- The medium should have a high transparency (low absorption and scattering) in the wavelength regions of pump and laser radiation, and good optical homogeneity. To some extent, this depends on the quality of the material, determined by details of the fabrication process.
- The host medium influences strongly the wavelength, bandwidth and transition cross sections of pump and laser transitions and also the upper-state lifetime. For example, Nd:YVO_4 has much higher cross sections, a larger gain bandwidth, and a smaller upper-state lifetime than Nd:YAG. Other neodymium hosts provide other transition wavelengths, e.g. 1047 or 1053 nm from Nd:YLF.
- Nonradiative transitions (e.g. multi-phonon transitions) are also strongly influenced by the host, in particular by its maximum phonon energy. Some of these transitions are very detrimental, leading to quenching of the upper-state population (thus lowering the quantum efficiency). Others are essential for laser operation, e.g. for removing ions from the lower laser level.

  Energy transfer processes are also dependent on the host material.
- The maximum possible doping concentration can depend strongly on the host material and its fabrication method.
- Different crystalline materials are very different concerning their hardness and other properties, which determine with which methods and how easily they can be cut and polished with good quality.

- Some materials are chemically not stable, e.g. hygroscopic.
- Particularly for high-power lasers (but often enough also for medium and low powers), a high thermal conductivity low thermo-optic coefficients (for weak thermal lensing) and a high resistance to mechanical stress are desirable.
- Optical isotropy can be beneficial, but in other cases birefringence (reducing thermal depolarization) and possibly polarization-dependent gain is preferable.
- A high damage threshold in terms of pulse fluence or peak intensity can be important for high-energy amplifiers.

It is apparent that different applications lead to very different requirements on laser gain media. For this reason, a broad range of different crystals are used, and making the right choice is essential for constructing lasers with optimum performance.

Common Crystalline Laser Host Media

There is a wide range of crystalline media, which can be grouped according to important atomic constituents and crystalline structures. Some important groups of crystals are:

- Garnets such as $Y_3Al_5O_{12}$ (YAG), $Gd_3Ga_5O_{12}$ (GGG), and $Gd_3Sc_2Al_3O_{12}$ (GSGG): hard and chemically inert materials, optical isotropic, with high thermal conductivity
- sapphire (Al_2O_3) (e.g. for titanium–sapphire lasers) and aluminates such as $YAlO_3$ (YALO, YAP) for neodymium doping: high hardness and thermal conductivity, anisotropic
- Sesquioxides such as Y_2O_3, Sc_2O_3: isotropic, high hardness and thermal conductivity
- Vanadates such as YVO_4 and $GdVO_4$: very high laser cross sections of Nd^{3+}, anisotropic
- fluorides, e.g. $YLiF_4$ (YLF): good UV transparency, birefringence, large energy storage capability of

Nd:YLF; also LiCAF, LiLuF, LiSAF as chromium-doped broadband gain media
- Silicates, e.g. $MgSiO_4$ (forsterite): broad gain bandwidth
- monoclinic double tungstates such as $KGd(WO_4)_2$ (KGW) and $KY(WO_4)_2$ (KYW): combination of relatively high Yb^{3+} laser cross sections, large gain bandwidth, and high thermal conductivity
- Disordered tetragonal double tungstates such as $NaGd(WO_4)_2$ (NGW) and $NaY(WO_4)_2$ (NYW): particularly large gain bandwidth of ytterbium
- Chalcogenides such as ZnS or ZnSe for mid-infrared lasers

Geometries of Laser Crystals

Different geometric forms can be used in lasers:

- A common form is that of a cuboid. The crystal can be, e.g., a thin coplanar plate, with transverse dimensions (perpendicular to the laser beam) and a thickness of a few millimeters. It may be oriented for near perpendicular incidence of the laser beam, or at Brewster's angle.
  It can be fixed in some solid mount which also acts as a heat sink. Larger crystals are usually used for side pumping e.g. with high-power diode bars.
- In some cases, extreme angles between the end faces are required, e.g. if one end face has to be Brewster-angled while the other one is made for perpendicular incidence.
- Slab lasers are based on relatively flat slabs, which may or may not be of cuboid form.
- Many side-pumped lasers use relatively long cylindrical laser rods, e.g. made of Nd:YAG. Particularly for lamp-pumped lasers, the rod length can be several centimeters, whereas the rod diameter is much smaller (a few millimeters).
- Thin-disk lasers require a disk, often with circular

cross section, having a thickness of only e.g. 100–200 ìm and a relatively high doping concentration.

- Special geometries are required for monolithic solid-state lasers, such as nonplanar ring oscillators.
- For various reasons, composite crystals are becoming popular. These have a spatially varying chemical composition and can be made with special shapes.

Bulk Properties

For a given dopant and host medium, the doping concentration is the most important parameter. Other issues of interest are the uniformity of doping (influencing the tendency for quenching), the level of impurities (e.g. unwanted other rare earth ions), and the optical homogeneities. Several of these factors influence the absorption and scattering losses of the material, and/or the strength of thermal lensing.

Of course, it is very desirable that a given crystal quality is produced consistently, although different laser designs can have a different sensitivity to material parameters.

Optimization of Geometry and Parameters

Which geometry, dopant and doping concentration of the gain medium are most advantageous depend on several factors. The available pump source (type of laser diode or lamp) and the envisaged pumping arrangement are important factors, but the material itself also has some influence.

For example, titanium–sapphire lasers have to be pumped with high intensities, for which the form of a transversely cooled rod, operated with relatively small pump and laser beam diameter, is more appropriate than e.g. a thin disk. As another example, Q-switched lasers reach a higher population density in the upper laser level and are therefore more sensitive to quenching effects and energy transfer processes; therefore, a lower doping density is often appropriate for these devices.

For high-power lasers, lower doping densities are often used in order to limit the density of heat generation, although thin-disk lasers work best with highly doped crystals. Many laser products do not reach the full performance potential because such details have not been properly worked out.

Optical Surfaces

Those surfaces which are passed by the laser beam are normally either oriented at Brewster's angle or have an anti-reflection coating. Even AR-coated crystals are often slightly tilted against the beam so as to prevent back-reflections staying in the laser resonator. This is important for, e.g., mode-locked lasers and tunable single-frequency lasers.

A high surface quality is of course important. Specifications of surface flatness are often better than λ/ 10. This helps to avoid both scattering losses and wavefront distortions which can degrade the laser's beam quality. In addition, *scratch and dig specifications* (*cosmetic surface quality*) limit the density of small-scale surface defects; they may read e.g. "80–50" for medium quality mass production, or "10–5" for particularly demanding laser applications. Proper surface treatment also influences the damage threshold, which is important e.g. for high-energy pulse amplifiers. Finally, a high degree of end face parallelism can be important for avoiding changes of beam direction in a crystal.

Diode-pumped Lasers

Virtually all optically pumped lasers fall into one of two categories:

- *Lamp-pumped lasers,* having some kind of discharge lamp as the pump source
- *Diode-pumped lasers,* pumped with some kind of laser diodes

The latter category, for which the term all-solid-state lasers is also used.

Types of Diode-pumped Lasers

Most diode-pumped lasers are solid-state lasers (DPSSL = DPSS lasers = *diode-pumped solid-state lasers*). These are either bulk lasers, using some kind of laser crystal or bulk piece of glass, or fiber lasers (although the term DPSSL is less common for fiber lasers). Both categories span a range of output powers from a few milliwatts to multiple kilowatts (→ *high-power lasers*).

Less common are optically pumped semiconductor lasers (particularly VECSELs = vertical external cavity surface-emitting lasers), and there are also some relatively exotic types of diode-pumped gas lasers, e.g. alkali vapor lasers.

Types of Laser Diodes for Diode Pumping

There are different types of laser diodes which can be used for diode pumping:

- Low-power lasers (up to roughly 200 mW) can be pumped with small edge-emitting laser diodes. These exhibit a diffraction-limited beam quality and make it fairly easy to achieve the same for the solid-state laser.
- Broad area laser diodes typically generate several watts and are suitable for pumping solid-state lasers with output powers up to a few watts. Their beam quality is substantially asymmetric, but normally still sufficient for achieving a diffraction-limited laser output without using complicated optics.
- High-power diode bars emit tens of watts (or even > 100 W), allowing for higher output powers, particularly when several bars are combined. Their beam quality is strongly asymmetric and fairly poor, so that their brightness (radiance) is much lower than that of lower-power diodes. Various types of beam shapers are used to symmetrize the beam quality. This makes it easier either to pump a bulk laser or to couple the light into a fiber.
- For the highest powers, diode stacks are often used. These have a still worse beam quality and lower brightness, but can provide multiple kilowatts.

In most cases, the pump diodes are operated continuously. This applies to all continuous-wave and mode-locked lasers, and also to many Q-switched lasers. However, quasi-continuous-wave operation with higher peak power for limited time intervals (e.g. 100 ìs) is sometimes used for Q-switched lasers with a high pulse energy and low pulse repetition rate.

Depending on the type of laser diode, different kinds of pump optics are used. It is also possible to use fiber-coupled diode lasers, which make it possible to separate the actual laser head from another package containing the pump diodes, so that the laser head can become very compact.

Advantages of Diode Pumping

The main advantages of diode pumping are:

- A high electrical-to-optical efficiency of the pump source (of the order of 50%) leads to a high overall power efficiency (→ *wall-plug efficiency*) of the laser. As a consequence, small power supplies are needed, and both the electricity consumption and the cooling demands are drastically reduced, compared with those for lamp-pumped lasers.
- The narrow optical bandwidth of diode lasers makes it possible to pump directly certain transitions of laser-active ions without losing power in other spectral regions. It thus also contributes to a high efficiency.
- Although the beam quality of high-power diode lasers is not perfect, it often allows for end pumping of lasers with very good overlap of laser mode and pump region, leading to high beam quality and power efficiency. In the domain of slab lasers, it allows edge pumping instead of face pumping, which brings important advantages.
- Diode-pumped low-power lasers can be pumped with diffraction-limited laser diodes. This allows the construction of very low-power lasers with reasonable power efficiency, i.e., with a very small electrical pump power, as is important for battery-powered devices.
- The lifetime of laser diodes is long compared with that of discharge lamps: typically many thousands of hours, often even well above 10 000 hours.

However, various factors can lead to drastically reduced lifetimes. Also, the exchange of laser diodes is much more costly than that of discharge lamps.

- The compactness of the pump source, the power supply and the cooling arrangement makes the whole laser system much smaller and easier to use.
- Diode pumping makes it possible to use a very wide range of solid-state gain media for different wavelength regions, including e.g. upconversion lasers. For many solid-state gain media, the lower brightness of discharge lamps would not be sufficient.
- The low intensity noise of laser diodes leads to low noise of the diode-pumped laser.

Achievements

The benefits of diode pumping have lead to amazing achievements. Some examples are:

- There are miniature solid-state lasers with excellent efficiency, beam quality, spectral purity, and stability. Some of those may even be battery-powered.
- Diode-pumped high-power solid-state lasers can deliver kilowatt output powers with fairly high beam quality. This applies particularly to thin-disk lasers, but also to high-power fiber lasers and amplifiers.
- Diode pumping is also essential for a large variety of mode-locked lasers, generating e.g. up to 80 W of average output power in sub-picosecond pulses, or picosecond pulses for telecom applications with a pulse repetition rate up to 50 GHz.

Limitations

In the early years of diode pumping, the output powers achievable were very limited – smaller than those of lamp-pumped lasers. In the meantime, however, high-power diode bars and diode stacks have become very powerful, and the highest output powers are now usually achieved with diode pumping.

The main disadvantage of diode pumping (as compared with lamp pumping) is the significantly higher cost per watt of pump power. This is severe for high powers. For this reason, lamp pumping is still used in cases where high powers are needed, particularly when the power is used only for short times. For example, lamp-pumped Q-switched Nd:YAG lasers are still widely used for laser marking, and will not soon be replaced with diode-pumped lasers.

Laser diodes are electrically less robust than discharge lamps. They may e.g. be quickly destroyed by excessive drive currents, or by electrostatic discharges. In conjunction with properly designed electronics, however, this should not happen. Problems can also arise from optical feedback.

Applications

Diode-pumped solid-state lasers have a very wide range of applications. Indeed, they are used in all of the areas mentioned in the article on laser applications.

FIBER LASERS

Fiber lasers are usually meant to be lasers with optical fibers as gain media, although some lasers with a semiconductor gain medium (a semiconductor optical amplifier) and a fiber resonator have also been called fiber lasers (or *semiconductor fiber lasers*). In most cases, the gain medium is a fiber doped with rare earth ions such as erbium (Er^{3+}), neodymium (Nd^{3+}), ytterbium (Yb^{3+}), thulium (Tm^{3+}), or praseodymium (Pr^{3+}), and one or several laser diodes are used for pumping (→ *diode pumping*).

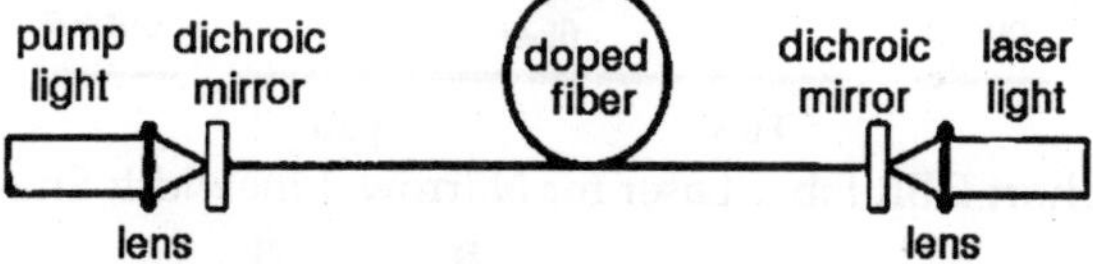

Fig. Setup of A Simple Fiber Laser. Pump Light is Launched from the Left-Hand Side Through a Dichroic Mirror into the Core of the Doped Fiber. The Generated Laser Light is Extracted on The Right-Hand Side.

FIBER LASER RESONATORS

In order to form a laser resonator with fibers, one either needs some kind of reflector (mirror) to form a linear resonator, or one builds a fiber ring laser. Various types of mirrors are used in linear fiber laser resonators:

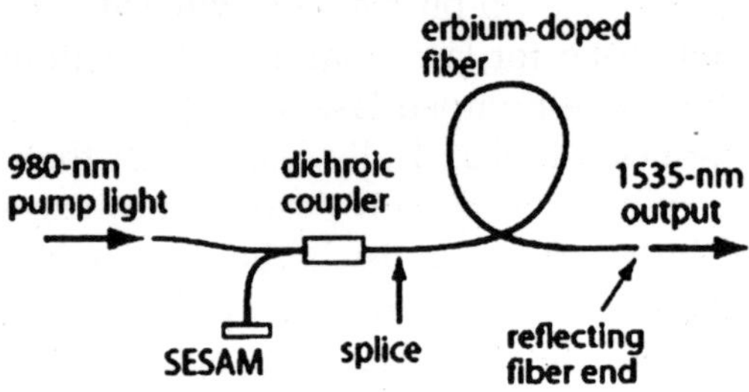

Fig. A Simple Erbium-Doped Femtosecond Laser, where the Fresnel Reflection from a Fiber end is used for Output Coupling.

- In simple laboratory setups, ordinary dielectric mirrors can be butted to the perpendicularly cleaved fiber ends. This approach, however, is not very practical for mass fabrication and not very durable either.
- The Fresnel reflection from a bare fiber end face is often sufficient for the output coupler of a fiber laser.
- It is also possible to deposit dielectric coatings directly on fiber ends, usually with some evaporation method. Such coatings can be used to realize reflectivities in a wide range.
- For commercial products, it is common to use fiber Bragg gratings, made either directly in the doped fiber, or in an undoped fiber which is spliced to the active fiber.

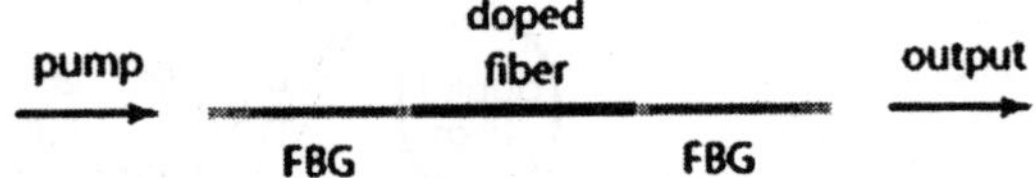

Fig. Short DBR Fiber Laser for Narrow–Linewidth Emission.

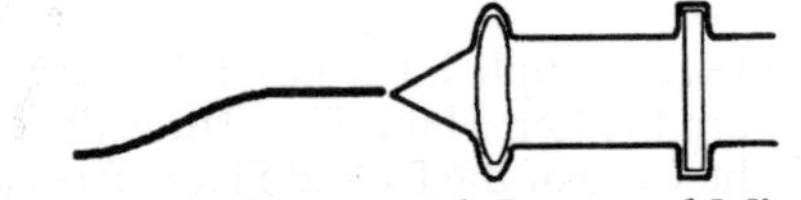

Fig. End Reflector with Lens and Mirror.

- A better power-handling capability is achieved by collimating the light exiting the fiber with a lens and reflecting it back with a dielectric mirror. The intensities on the mirror are then greatly reduced due to the much larger beam area. However, slight misalignment can cause substantial reflection losses, and the additional Fresnel reflection at the fiber end can introduce filter effects and the like. The latter effects can be suppressed by using angle-cleaved fiber ends, which however introduce polarization-dependent losses.

Fig. Fiber Loop Mirror.

- Another option is to form a fiber loop mirror, based on a fiber coupler (e.g. with 50:50 splitting ratio) and some piece of passive fiber.

Most fiber lasers are pumped with one or several fiber-coupled diode lasers. The pump light may be coupled directly into the core, or in high-power into a larger pump cladding (→ *double-clad fibers*).

There are many different kinds of fiber lasers, some of which are discussed in the following.

High-power Fiber Lasers

Whereas the first fiber lasers could deliver only a few milliwatts of output power, there are now high-power fiber lasers with output powers of hundreds of watts, sometimes even several kilowatts from a single fiber. This potential arises from a high surface-to-volume ratio (avoiding excessive heating) and the guiding effect, which avoids thermo-optical problems even under conditions of significant heating.

Upconversion Fiber Lasers

The fiber laser concept is most suitable for the realization of upconversion lasers, as these often have to operate on relatively "difficult" laser transitions, requiring high pump intensities. In a fiber laser, such high pump intensities can be easily maintained over a long length, so that the gain efficiency achievable often makes it easy to operate even on low-gain transitions.

In most cases, silica glass is not suitable for upconversion fiber lasers, because the upconversion scheme requires relatively long lifetimes of intermediate electronic levels, and such lifetimes are often very small in silica fibers due to the relatively large phonon energy of silica glass (→ *multi-phonon transitions*). Therefore, one mostly uses certain heavy-metal fluoride fibers such as ZBLAN (a fluorozirconate) with low phonon energies.

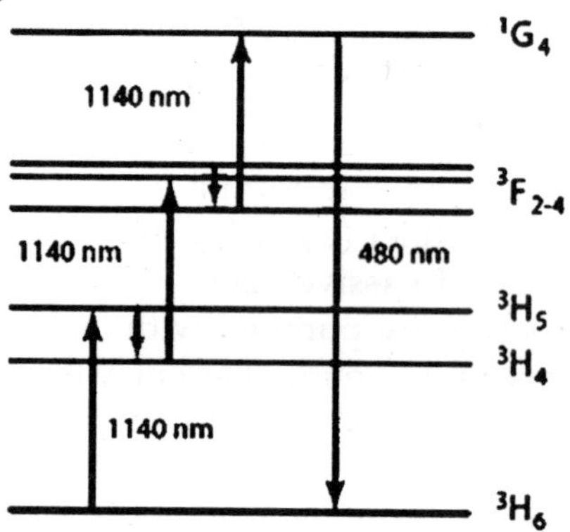

Fig. Level Scheme of Thulium (Tm^{3+}) ions in ZBLAN Fiber

The probably most popular upconversion fiber lasers are based on thulium-doped fibers for blue light generation, praseodymium-doped lasers (possibly with ytterbium codoping) for red, orange, green or blue output, and green erbium-doped lasers.

Narrow-linewidth Fiber Lasers

Fiber lasers can be constructed to operate on a single longitudinal mode (→ *single-frequency lasers, single-mode operation*) with a very narrow linewidth of a few kilohertz or even below 1 kHz. In order to achieve long-term stable single-frequency operation without excessive requirements concerning temperature stability, one usually has to keep the laser resonator relatively short (e.g. of the order of 5 cm), even though a longer resonator would in principle allow for even lower phase noise and a correspondingly smaller linewidth.

The fiber ends have narrow-bandwidth fiber Bragg gratings (→ *distributed Bragg reflector lasers, DBR fiber lasers*), selecting a single resonator mode. Typical output powers are a few milliwatts to some tens of milliwatts, although single-frequency fiber lasers with up to roughly 1 W output power have also been demonstrated.

An extreme form is the distributed-feedback laser (DFB laser), where the whole laser resonator is contained in a fiber Bragg grating with a phase shift in the middle. Here, the resonator is fairly short, which can compromise the output power and linewidth, but single-frequency operation is very stable.

Of course, further amplification to much higher power levels in a fiber amplifier is possible.

Q-switched Fiber Lasers

With various methods of active or passive Q switching, fiber lasers can be used for generating pulses with durations which are typically between tens and hundreds of nanoseconds. The pulse energy achievable with large mode area fibers can be several millijoules, in extreme cases tens of millijoules, and is essentially limited by the saturation energy (even for large mode area fibers) and by the damage threshold (the latter particularly for shorter pulses). As fiber lasers typically have relatively long resonators (particularly for high-power lasers based on double-clad fibers), the pulse durations tend to be longer than those of bulk lasers.

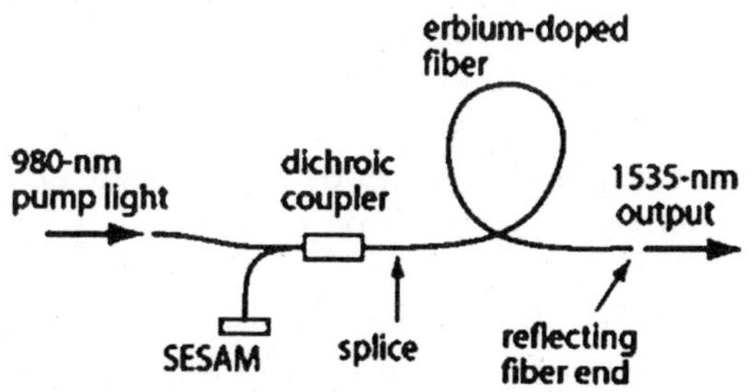

Fig. Simple Erbium-Doped Q-Switched Fiber Laser.

Mode-locked Fiber Lasers

More sophisticated resonator setups are used particularly for mode-locked fiber lasers (*ultrafast fiber lasers*), generating picosecond or femtosecond pulses. Here, the laser resonator may contain an active modulator or some kind of saturable absorber.

An artificial saturable absorber can be constructed using the effect of nonlinear polarization rotation, or a nonlinear fiber loop mirror. A nonlinear loop mirror is used e.g. in a "figure-of-eight laser" where there is a main resonator on the left-hand side and a nonlinear fiber loop, which does the amplification, shaping and stabilization of a circulating ultrashort pulse. Particularly for harmonic mode locking, additional means may be used, such as subcavities acting as optical filters.

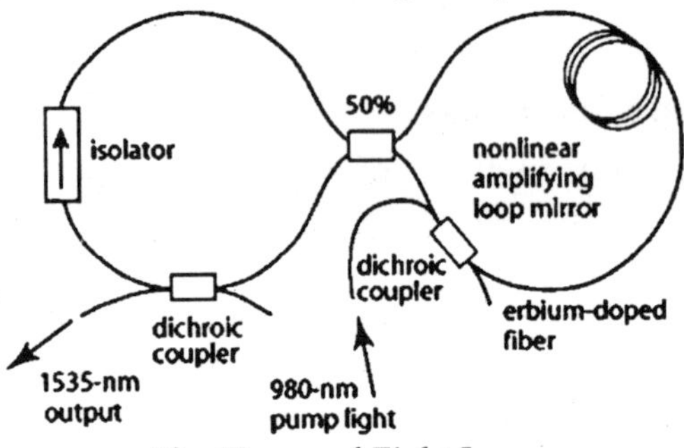

Fig. Figure-of-Eight Laser

RAMAN FIBER LASERS

A special type of fiber lasers are fiber Raman lasers, relying on Raman gain associated with the fiber nonlinearity. Such lasers usually use relatively long fibers, sometimes of a type with increased nonlinearity, and typical pump powers of the order of 1W.

With several nested pairs of fiber Bragg gratings, the Raman conversion can be done in several steps, bridging hundreds of nanometers between the pump and output wavelength. Raman fiber lasers can e.g. be pumped in the 1-ìm region and generate 1.4-ìm light as required for pumping 1.5-ìm erbium-doped fiber amplifiers.

Fiber Lasers with Semiconductor Optical Amplifiers

There are some lasers which have a semiconductor optical amplifier (SOA) as the gain medium in a resonator made of fibers.

Even though the actual laser process does not occur in a fiber, such fibers are sometimes called fiber lasers. They typically emit relatively small optical powers of a few milliwatts or even less.

Sometimes they exploit the very different properties of the semiconductor gain medium, as compared with a rare-earth-doped fiber, in particular the much smaller saturation energy and upper-state lifetime.

Rather than only generating coherent light, such lasers can be used for information processing in optical fiber communications systems – for example the wavelength conversion of data channels based on cross-saturation effects.

Special Attractions of Fibers as Laser Gain Media

- As fibers can be coiled and the light propagating in fibers is well shielded from the environment (e.g. concerning dust), fiber lasers can have a compact and rugged setup, provided that the whole laser resonator is built only with fiber components (*all-fiber setup*) such as fiber Bragg gratings and fiber couplers (i.e., avoiding free-space optics and any requirement for alignment).
- Fiber gain media have a large gain bandwidth due to strongly broadened laser transitions in glasses, permitting wide wavelength tuning ranges and/or the generation of ultrashort pulses. Also, fiber lasers have broad spectral regions with good pump absorption, making the exact pump wavelength uncritical, so that temperature stabilization of the pump diodes is usually not necessary.
- Diffraction-limited beam quality is easily obtained when single-mode fibers are used, and sometimes also with slightly multimode fibers.

- Due to the high gain efficiency of doped fibers, fiber lasers have the potential to operate with very small pump powers. Also, it is possible to obtain very high power efficiencies.
- In recent years, the potential for very high output powers (several kilowatts with double-clad fibers) has been convincingly demonstrated.
- Again due to the guidance, which allows high pump intensities to be applied over long lengths, fiber lasers can be operated even on very "difficult" laser transitions (e.g. of upconversion lasers).

On the other hand, fiber lasers can suffer from various problems:

- When the pump light has to be launched from free space into a single-mode core, the alignment is critical. This problem can be eliminated by using fiber-coupled pump diodes.
- Most fibers exhibit a complicated temperature-dependent polarization evolution, unless polarization-maintaining fibers or Faraday rotators are used. Such measures, however, are normally not compatible with nonlinear polarization rotation mode locking.
- Nonlinear effects often limit the performance, e.g. in terms of powers achievable in single-frequency operation or the pulse quality of mode-locked lasers. For example, Kelly sidebands are often seen, whereas mode-locked bulk lasers rarely exhibit this effect.
- At high powers, there is a risk of fiber damage even below the actual damage threshold of the material (→ *fiber fuse*).
- Fibers have a limited gain and pump absorption per unit length, making it difficult to realize very short resonators e.g. for single-frequency lasers or for multi-gigahertz mode-locked lasers. However, significant progress has been made in this direction recently via the development of very highly doped fibers, usually made from phosphate glass.

Double-clad Fibers

A fiber laser or amplifier based on an ordinary doped single-mode fiber can generate a diffraction-limited output, but it restricts the pump sources to those with diffraction-limited beam quality and thus normally to those with low power.

On the other hand, the use of multimode fibers usually (although not always) leads to poor beam quality. This dilemma has been resolved with the invention of double-clad fiber designs, which allow *cladding pumping of fiber devices*. Here, the laser light propagates in a single-mode (or multimode) core, which is surrounded by an *inner cladding* in which the pump light propagates. Only the core (or sometimes a ring around the core) is rare-earth-doped.

The pump light is restricted to the inner cladding by an outer cladding with lower refractive index, and also partly propagates in the single-mode core, where it can be absorbed by the laser-active ions. The inner cladding has a significantly larger area (compared with that of the core) and typically a much higher numerical aperture, so that it can support a large number of propagation modes, allowing the efficient launch of the output, e.g. of high-power laser diodes (e.g. beam-shaped high-power diode bars), despite their poor beam quality.

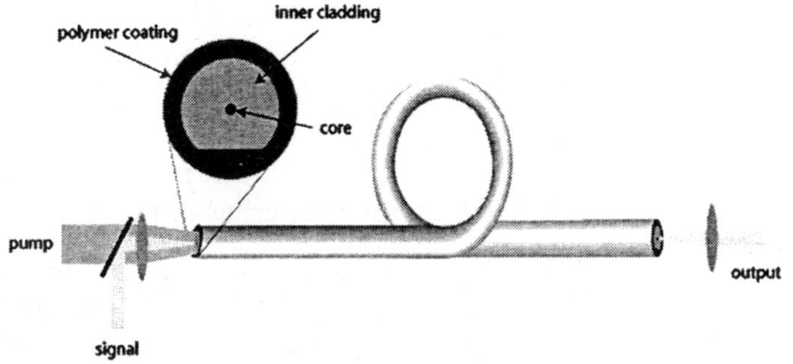

Fig. Cladding-Pumped Fiber
Amplifier based on a Double-Clad Fiber.

The signal light is launched into the doped core, while the pump light is launched into the inner cladding. The core is D-shaped for more efficient pump absorption. The pump light does not necessarily need to be injected into the fiber ends. It is also possible to use side pumping techniques, where access to the fiber ends is not required for pumping. For example, coated V grooves cut into the inner cladding can be used to reflect pump light into the inner cladding.

Double-clad Fiber Designs

There are a variety of different designs of double-clad fibers. The fiber cross-sections for the most important design types. The simplest kind of design has a circular pump cladding and a centered core.

This is relatively easy to make and use, but in this kind of fibers there are propagation modes of the inner cladding (related to *helical rays*) which have hardly any overlap with the core, so that some significant part of the pump light exhibits incomplete absorption. As a result, the gain and power efficiency are compromised. To a limited extent, this problem can be solved by strongly coiling the fiber.

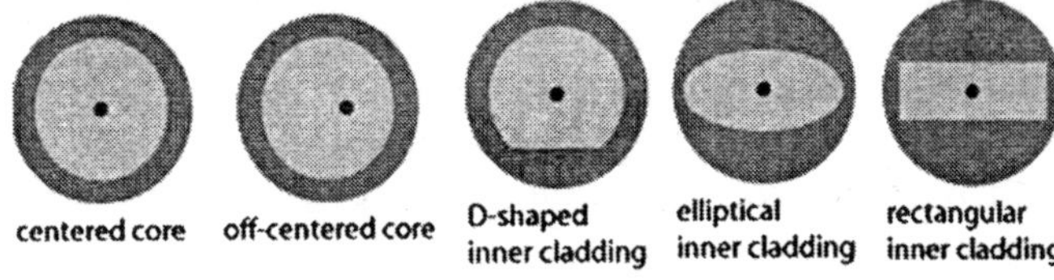

Fig. Various Designs of Double-Clad Fibers.

Modes with poor core overlap can be avoided by using a modified design with a lower symmetry. Examples are designs with an off-centered core or a non-circular (e.g. elliptical, D-shaped or rectangular) inner cladding. Such pump claddings are also often better matching the properties of pump sources such as beam-shaped diode bars.

Here, the multimode pump core is suspended by very thin struts in the air cladding, through which the pump light cannot escape. Such a structure can have a very high numerical aperture of at least 0.6 for the pump light; this further reduces the requirements concerning the brightness of the pump source. The thickness of the struts can be chosen such that at the same time one achieves good mechanical stability, high thermal conductivity, and minimal pump losses.

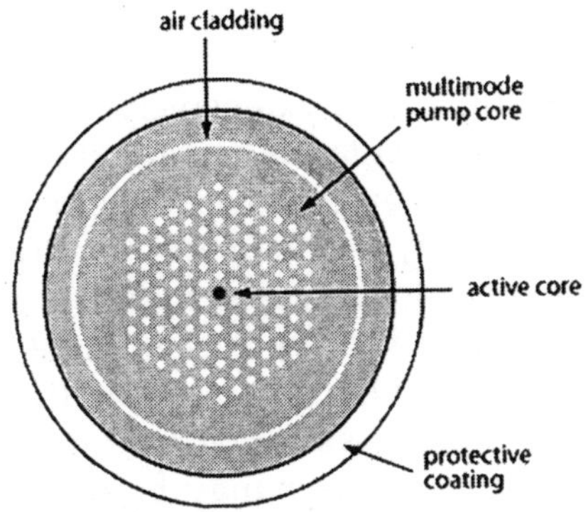

Fig. Structure of a Photonic Crystal Fiber with an Air Cladding.

Another advantage of this type of fiber is that pump light is kept away from the protective polymer coating, avoiding any damage by absorbed pump light. The guidance of the core is achieved as in other photonic crystal fibers.

Parameters and Fabrication Methods of Double-clad Fibers

Besides the properties of the fiber core, the ratio of the areas of inner cladding and core is an important parameter. This area ratio should not be too large, because otherwise the effective pump absorption length becomes large, and the pump intensity in the core is small, resulting in low excitation levels which can also compromise the power efficiency. Area ratios of the order of 100–1000 are common. Pump sources with improved brightness allow the use of fibers with a smaller area ratio, and thus also with a smaller length, which also reduces the impact of various types of nonlinearities.

In many cases, the core and inner cladding of a double-clad fiber are similar to those of a normal core-pumped fiber, except that in addition there is the lower-index outer cladding. If the inner cladding is made of silica, the outer cladding may consist of fluorine-doped silica.

The numerical aperture for the inner cladding can then be e.g. 0.28. Larger values are possible with polymer outer claddings, but these cannot tolerate very high temperatures and may introduce higher propagation losses for the pump light.

Applications

Double-clad fibers are extensively used for cladding-pumped high-power fiber lasers and amplifiers. Such devices can have a fairly high power conversion efficiency combined with good beam quality. As the beam quality of the output can be diffraction-limited whereas that of the pump can be poor, the brightness of the laser or amplifier output can be much higher than that of the pump source. Particularly if this increase in brightness is essential for an application, the cladding-pumped fiber laser may be called a *brightness converter*.

High-power Fiber Lasers and Amplifiers

Whereas the first fiber lasers could deliver only a few milliwatts of output power, there have subsequently been rapid developments which have lead to high-power fiber lasers and particularly amplifiers with output powers of tens or hundreds of watts, sometimes even several kilowatts from a single fiber. This potential arises from a very high surface-to-volume ratio (avoiding excessive heating) and the guiding (waveguide) effect, which avoids thermo-optical problems even under conditions of significant heating.

Fiber laser technology now competes strongly with other high-power laser technologies based on solid-state bulk lasers, such as thin-disk lasers.

Double-clad Fibers and Beam Quality

High-power fiber lasers and fiber amplifiers are nearly always realized with rare-earth-doped double-clad fibers, which are pumped with fiber-coupled high-power diode bars or other kinds of laser diodes. The pump light is launched into an inner cladding rather than into the (much smaller) fiber core, in which the laser light is generated. The laser light can have very good beam quality – even diffraction-limited beam quality if the fiber has a single-mode core. Therefore, the brightness of the fiber laser output can be orders of magnitude higher than that of the pump light, even though the output power is of course somewhat smaller. (Typical power efficiencies are above 50%, sometimes even above 80%.) Such fiber lasers can effectively be used as brightness converters, i.e., devices increasing the brightness.

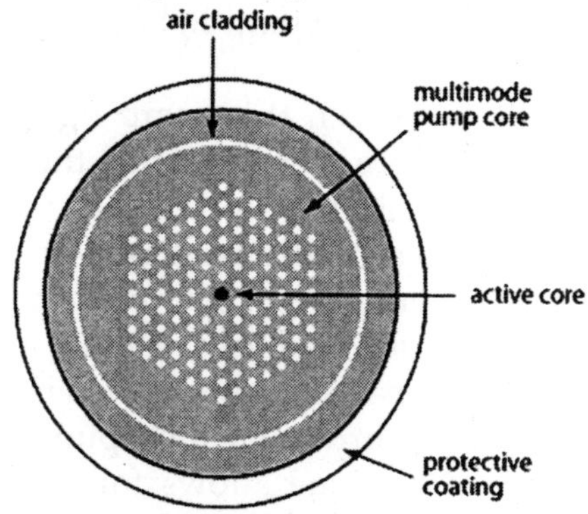

Fig. Structure of a Modern Double-Clad Fiber with an Air Cladding.

For the highest powers, the core area needs to be fairly large (→ *large mode area fibers*), because the optical intensities would otherwise become too high, and often also because a double-clad fiber with a large ratio of cladding to core area has a weak pump absorption. For core areas up to the order of a few thousand μm^2, it is feasible to have a single-mode core. Larger mode areas with still fairly good output beam quality are possible with a slightly multimode core, where most of the light propagates in the fundamental mode. (The excitation of higher-order fiber modes can be suppressed to some extent e.g. by coiling the fiber.) For even larger mode areas, the beam quality can no longer be nearly diffraction-limited, but it can still be fairly good compared with, e.g., rod lasers operating at similar power levels.

LAUNCHING THE PUMP LIGHT

There are several options for launching pump light at very high power levels. The simplest is to launch directly into the pump cladding at one or both fiber ends. This technique does not require special fiber components; however, it needs the propagation of the high-power pump radiation through air (with free-space optics) and particularly through an air–glass interface, which is then very sensitive to dust and misalignment. In many cases, it is therefore preferable to use one of several techniques where one employs fiber-coupled pump diodes and keeps the pump light in fibers from there on.

One option is to launch the pump light into passive (undoped) fibers which are wound around the active fiber so that the light is gradually transferred into the active fiber (*GTWave fiber*). Other techniques are based on special pump combiner devices, where several pump fibers and a single active signal fiber are fused together. Yet other approaches are based on side-pumped fiber coils (*fiber disk lasers*) or on grooves in the pump cladding through which pump light can be injected. The latter technique allows for *multi-point pump injection* and thus for a better distribution of the heat load.

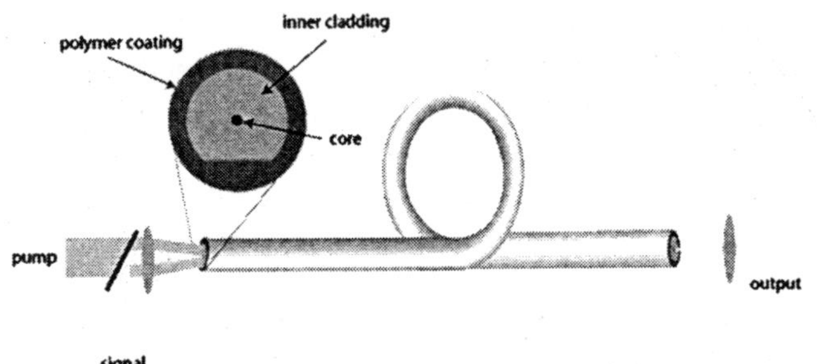

Fig. A High-power Double-Vlad Fiber

A comparison of all the different pump launching techniques is complex, since it involves many aspects: not only the transfer efficiency, but also the loss of brightness, the ease of manufacturing, the flexibility of handling, possible back-reflections, leakage of light from the fiber core back to the pump source, the option to maintain the polarization, etc.

LASERS AND AMPLIFIERS

The conceptually simplest option for generating high-power laser light is to build a fiber laser directly with some kind of mirrors on both ends. However, high-power fiber devices are very often built as laser-amplifier combinations, i.e., with a MOFA (*master oscillator fiber amplifier*) architecture. This concept has several advantages. It is easier to control the emission properties of a lower-power seed laser in terms of linewidth, laser noise, wavelength tunability, pulse generation, etc.

Also, the fiber of an amplifier has to stand only a power about equal to the output power, whereas in a laser the intracavity power is higher (even though fiber lasers as high-gain devices allow for strong output coupling). Furthermore, it can be advantageous to use a modular design approach, where amplifier stages can be combined as required. In many cases, particularly when a low-power seed laser is used, one even uses several amplifier stages, typically with increasing mode areas and pump powers along the chain.

NANOSECOND PULSES

Many high-power lasers for material processing are Q-switched lasers, generating intense nanosecond pulses. In this regime, fiber devices are limited mainly in terms of peak power: apart from various nonlinearities, self-focusing introduces a hard limit at a few megawatts of peak power, which notably cannot be increased by using larger mode areas. Actually achieved peak powers are usually below 1 MW, even for large mode area fibers. This means that, e.g., pulse energies of a few millijoules can not be combined with pulse durations of a few nanoseconds only.

On the other hand, the high gain of fiber devices allows the realization of very flexible MOPA devices with a gain-switched laser diode as seed laser. This approach, unlike Q switching of a laser, allows one, e.g., to modify the pulse duration independently of the pulse repetition rate.

ULTRASHORT PULSES

The large gain bandwidth of fiber amplifiers allows for the amplification of ultrashort pulses. However, particularly in this regime significant challenges arise from the strong fiber nonlinearity, because high-energy femtosecond pulses have huge peak powers. In addition to the risk of fiber damage, strong nonlinear distortions can result from the fiber nonlinearity, and the high level of chromatic dispersion (including higher-order dispersion) can also be problematic. Particularly if pulse quality matters, these matters need to be carefully considered.

One possible approach is to make fiber-based chirped-pulse amplification systems, where the pulse duration within the amplifier is strongly increased, so that the effect of nonlinearities is accordingly decreased. An alternative to this approach is the amplification of parabolic pulses, where up-chirped pulses experiencing the gain and nonlinearity of the amplifier fiber evolve in a self-similar fashion, and their close to linear chirp makes it possible to obtain a high pulse quality with dispersive compression.

Although the mentioned techniques have allowed for extraordinary performance of fiber-based high-power ultrashort pulse sources (at least for multi-megahertz pulse repetition rates and correspondingly moderate pulse energies), these achievements are usually based on setups which involve a lot of free-space optics and hence lose many of the advantages of fiber systems.

An interesting direction is the development of all-fiber ultrashort pulse sources, possibly allowing for a final free-space pulse compressor (e.g. a transmission grating), but eliminating the need to launch pulses from free space into a fiber.

Prospects for Further Improvements

Even though the progress in the development of high-power fiber devices has been tremendous in recent years, various kinds of more or less severe limitations are now encountered, which are expected to slow this progress:

- The optical intensities in high-power fiber devices have been enormously increased. Now they are often close to the damage threshold of the material. Therefore, increased mode areas (→ *large mode area fibers*) are used, but the limits of this approach have also been more or less reached, at least if very high beam quality is required.

- The power dissipation per unit length has reached values of the order of 100 W/m, which causes significant heating for air-cooled fibers. However, water cooling will allow for significant further increases in power. A longer fiber with lower doping concentration makes the cooling easier, but increases the effect of nonlinearities.
- Fiber nonlinearities can be disturbing in various ways. Even in continuous-wave devices, the Raman gain can become so high (tens of decibels) that a significant part of the power is transferred to a longer-wavelength Stokes wave, which does not experience laser gain. Single-frequency operation is even much more limited due to stimulated Brillouin scattering, although a number of measures are known to mitigate this effect to some extent. In devices for ultrashort pulses from mode-locked lasers, self-phase modulation can cause strong spectral broadening. In addition, there can be other problems such as nonlinear polarization rotation.

Due to such limitations, high-power fiber devices can – contrary to common opinions – usually not be considered as power-scalable in a strict sense, at least not well beyond the already achieved performance level. This has important consequences e.g. when fiber laser technology is compared with that of thin-disk lasers.

Of course, even without true power scalability, a lot can be done to improve various kinds of high-power fiber devices. An important issue is the further development of advanced fiber designs, e.g. with very large mode areas and single-mode guidance, as often realized with photonic crystal fibers. Various kinds of fiber devices are also important, e.g. special pump couplers, tapers for connecting fibers with different mode sizes and special cooling arrangements. Once the limit of the power per fiber has been reached, beam combining will be a further option – fiber devices appear suitable for this technique.

Also for ultrashort pulse amplifier systems there are various interesting options to mitigate and partially even to exploit the fiber nonlinearity, e.g. for further spectral broadening and subsequent pulse compression.

Beam Quality

The beam quality of a laser beam can be defined in different ways, but is essentially a measure of how tightly a laser beam can be focused under certain conditions (e.g. with a limited beam divergence). The most common ways to quantify the beam quality are:

- Tthe beam parameter product (BPP), i.e., the product of beam radius at the beam waist with the far-field beam divergence angle
- Tthe M^2 factor, defined as the beam parameter product divided by the corresponding product for a diffraction-limited Gaussian beam with the same wavelength
- Tthe inverse M^2 factor, which is high (ideally 1) for beams with high beam quality

The best possible beam quality is achieved for a diffraction-limited Gaussian beam, having $M^2 = 1$. This value is closely approached by many lasers, in particular by solid-state bulk lasers operating on a single transverse mode (→ *single-mode operation*) and by fiber lasers based on single-mode fibers, also by some low-power laser diodes (particularly VCSELs).

On the other hand, in particular some high-power lasers (e.g. solid-state bulk lasers and semiconductor lasers such as diode bars) can have a very large M^2 of more than 100 or even well above 1000.

In solid-state lasers, this is often a result of thermally induced wavefront distortions in the gain medium and/or a mismatch of effective mode area and pumped area in the laser crystal, whereas in high-power semiconductor lasers the poor beam quality results from operation with a highly multimode waveguide. In both cases, the poor beam quality is associated with the excitation of higher-order resonator modes.

In the focus of a diffraction-limited beam (i.e., at the location where the beam radius reaches its minimum), the optical wavefronts are flat. Any scrambling of the wavefronts, e.g. due to optical components with poor quality, spherical aberrations of lenses, thermal effects in a gain medium, diffraction at apertures, or by parasitic reflections, can spoil the beam quality.

For monochromatic beams, the beam quality could in principle be restored e.g. with a phase mask which exactly compensates the wavefront distortions, but this is usually difficult in practice, even in cases where the distortions are stationary. A more flexible approach is to use adaptive optics in combination with a wavefront sensor.

It is possible to some extent to improve the beam quality of a laser beam with a nonresonant mode cleaner or a mode cleaner cavity. This, however, leads to some loss of optical power.

The brightness of a laser is determined by its output power together with its beam quality.

Note that the term *beam quality* is sometimes used with a qualitative meaning which has little to do with the focusability. For some applications, it is vital to obtain a smooth beam intensity profile, e.g. of Gaussian shape, whereas the beam divergence is not of interest. The "quality" of a laser beam may then not be characterized e.g. with an M^2: one beam may have a relatively small M^2 value but a multi-peaked beam profile, whereas another beam may have a smooth beam shape but a high divergence and thus a large M^2 value.

MEASUREMENT OF BEAM QUALITY

According to ISO Standard 11146, the beam quality factor M^2 can be calculated with a fitting procedure, applied to the measured evolution of the beam radius along the propagation direction (the so-called *caustic*). For correct results, a number of rules have to be observed, e.g. concerning the exact definition of the beam radius and the placing of data points.

There are commercially available beam profilers which can automatically perform beam quality measurements within a few seconds.

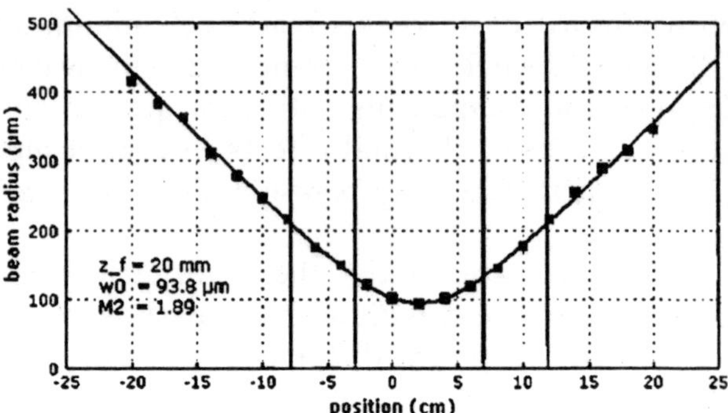

Fig. Calculation of the Beam Quality from the Measured Caustic.

They are normally based on the measurement of the beam profile at different positions. Beam profilers based on different measurement principles, e.g. CCD and CMOS cameras or rotating knife edges or slits, differ considerably in terms of the allowed ranges of beam radius and optical power, wavelength range, sensitivity to artifacts, etc.

For example, slit or knife-edge scanners can usually handle higher powers than cameras and can be precise for nearly Gaussian-shaped beams, whereas camera-based systems are usually more appropriate for complicated beam shapes. Other issues come into play for beams with temporally varying powers, e.g. for the output of Q-switched lasers. It may then be necessary to synchronize a shutter with the laser pulses.

Alternative measurement methods are based on the transmission through a mode-matched passive optical resonator or on wavefront sensors, e.g. Shack–Hartmann sensors. The full characterization of the laser beam then only requires analysis in a single plane.

Importance of Beam Quality for Applications

A high beam quality can be important e.g. when strong focusing of a beam is required. In the area of laser material processing, printing, marking, cutting and drilling require high beam qualities, whereas welding and various kinds of surface treatment are less critical in this respect, because they work with larger spots, so that direct application of high-power laser diodes with poor beam quality is possible.

For cutting and remote welding, a relatively high beam quality (with M^2 not much larger than 10) makes it possible to use a large working distance (i.e., a large distance between workpiece and focusing objective), which is highly desirable e.g. in order to protect the optics against debris and fumes. Also, a high beam quality reduces the beam diameters in a beam delivery system, so that smaller and thus cheaper optical elements (e.g. mirrors and lenses) can be used. Furthermore, the increased effective Rayleigh length (for a given spot size) increases the tolerance for longitudinal alignment.

A large working distance, made possible by a high beam quality, is also important for the design of diode-pumped lasers when the pump beam has to go through various pieces of optics (e.g. a dichroic mirror) before reaching the laser crystal.

A very high (close to diffraction-limited) beam quality, associated with a high spatial coherence, is often required for interferometers, optical data recording, laser microscopy, and the like.

Mode-locked lasers always have to have a high beam quality, since the excitation of higher-order transverse modes would disturb the pulse formation process.

Optimizing Laser Beam Quality

Crucial factors for obtaining a high beam quality from a solid-state bulk laser are:

- An optimized resonator design with suitable mode area (particularly in the gain medium) and low sensitivity to thermal lensing
- Good resonator alignment
- Minimized thermal effects, particularly from thermal lensing in the gain medium
- High-quality optical components (particularly concerning the gain medium)

- An optimized pump intensity distribution (sometimes requiring a pump source with good beam quality) – more easily achieved with end pumping than with side pumping

Beam Quality in Nonlinear Optics

Beam quality is an issue not only for lasers, but also for nonlinear frequency conversion. While thermal lensing in nonlinear crystal materials occurs only at very high average power levels (because heating occurs only through weak parasitic absorption), the beam quality can be affected by other effects:

- Spatial walk-off can spatially shift the interacting beams, so that the overlap becomes weaker, and the interaction becomes spatially asymmetric.
- For strong conversion e.g. in a frequency doubler or an optical parametric amplifier, there can be strong depletion of the pump beam near the beam axis or even backconversion, in extreme cases leading to pronounced ring structures. Gain guiding can make such problems more severe. Beam quality issues have been shown to limit the power scalability of high-gain nonlinear frequency conversion devices .
- For ultrashort pulses, the group velocity mismatch and other effects can even lead to time-dependent beam quality.

Further, the use of a laser beam with poor beam quality in a nonlinear frequency conversion device can significantly spoil the conversion efficiency.

Beam quality effects in nonlinear optics can be investigated with numerical computer models, which can simulate the evolution of the spatial (and possibly temporal) profiles of the involved beams.

RUBY LASER

A ruby laser is a solid-state laser that uses a synthetic ruby crystal as its gain medium. It was the first type of laser invented, and was first operated by Theodore H. "Ted"

Maiman at Hughes Research Laboratories on 1960-05-16.The ruby laser produces pulses of visible light at a wavelength of 694.3 nm, which appears as deep red to human eyes.

Typical ruby laser pulse lengths are on the order of a millisecond. These short pulses of red light are visible to the human eye, if the viewer carefully watches the target area where the pulse will fire.

Solid-state lasers are lasers based on solid-state gain media such as crystals or glasses doped with rare earth or transition metal ions, or semiconductor lasers. (Although semiconductor lasers are of course also solid-state devices, they are often not included in the term solid-state lasers.) Ion-doped solid-state lasers (also sometimes called *doped insulator lasers*) can be made in the form of bulk lasers, fiber lasers, or other types of waveguide lasers. Solid-state lasers may generate output powers between a few milliwatts and (in high-power versions) many kilowatts.

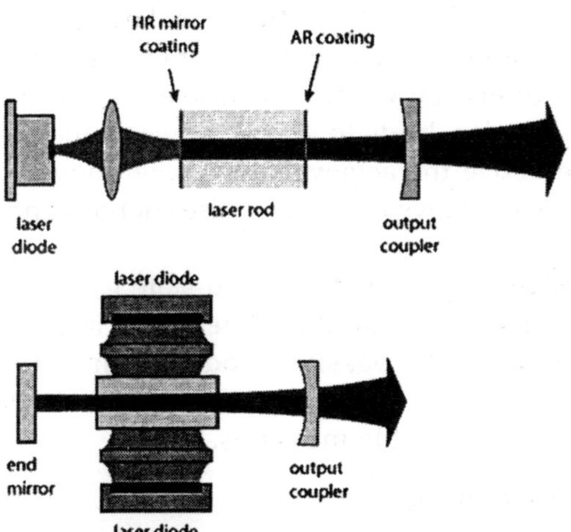

Fig. Typical Setups of Solid-State Bulk Lasers.

Optical Pumping and Energy Storage

Many solid-state lasers are optically pumped with flash lamps or arc lamps. Such pump sources are relatively cheap and can provide very high powers. However, they lead to a fairly low power efficiency, moderate lifetime, and strong thermal effects such as thermal lensing in the gain medium.

For such reasons, laser diodes are very often used for pumping solid-state lasers. Such *diode-pumped solid-state lasers* (DPSS lasers, also called *all-solid-state lasers*) have many advantages, in particular a compact setup, long lifetime, and often very good beam quality. Therefore, their share of the market is rapidly rising.

The laser transitions of rare-earth or transition-metal-doped crystals or glasses are normally weakly allowed transitions, i.e., transitions with very low oscillator strength, which leads to long radiative upper-state lifetimes and consequently to good energy storage, with upper-state lifetimes of microseconds to milliseconds. Although this is beneficial for nanosecond pulse generation, it can also lead to unwanted spiking phenomena in continuous-wave lasers, e.g. when the pump source is switched on.

Pulse Generation

The long upper-state lifetimes makes solid-state lasers very suitable for Q switching: the laser crystal can easily store an amount of energy which, when released in the form of a nanosecond pulse, leads to a peak power which is orders of magnitude above the achievable average power. Bulk lasers can thus easily achieve millijoule pulse energies and megawatt peak powers.

In mode-locked operation, solid-state lasers can generate ultrashort pulses with durations measured in picoseconds or femtoseconds. With passive mode locking, they have a tendency for Q-switching instabilities, if these are not suppressed with suitable measures.

Wavelength Tuning

In terms of their potential for wavelength tuning, different types of solid-state lasers differ considerably. Most rare-earth-doped laser crystals, such as Nd:YAG and Nd:YVO_4, have a fairly small gain bandwidth of the order of 1 nm or less, so that tuning is possible only within a rather limited range. On the other hand, tuning ranges of tens of nanometers and more are possible with rare-earth-doped glasses, and particularly with transition-metal-doped crystals such as Ti:sapphire, Cr:LiSAF and Cr:ZnSe (→ *vibronic lasers*).

Types of Solid-State Lasers

Examples of different types of solid-state lasers are:

- Small diode-pumped Nd:YAG (→ *YAG lasers*) or Nd:YVO_4 lasers (→ *vanadate lasers*) often operate with output powers between a few milliwatts (for miniature setups) and a few watts. Q-switched versions generate pulses with durations of a few nanoseconds, microjoule pulse energies and peak powers of many kilowatts. Intracavity frequency doubling can be used for green output.
- Single-frequency operation, typically achieved with unidirectional ring lasers (e.g. NPROs = nonplanar ring oscillators) or microchip lasers, allows for operation with very small linewidth in the lower kilohertz region.
- Larger lasers in side-pumped or end-pumped configurations, having the geometry of rod lasers, slab lasers or thin-disk lasers, are suitable for output powers up to several kilowatts. Particularly thin-disk lasers can still offer very high beam quality, and also a high power efficiency.
- Q-switched Nd:YAG lasers are still widely used in lamp-pumped versions. Pulsed pumping allows for high pulse energies, whereas the average output powers are often moderate (e.g. a few watts). The cost of such lamp-pumped lasers is lower than for diode-pumped versions with similar output powers.
- Fiber lasers are a special kind of solid-state lasers, with a high potential for high average output power, high power efficiency, high beam quality, and broad wavelength tunability.

Applications

Ruby lasers have declined in use with the discovery of better lasing media. They are still used in a number of applications where short pulses of red light are required. Holographers around the world produce holographic portraits with ruby lasers, in sizes up to a metre squared.

The red 694 nm laser light is preferred to the 532 nm green light of frequency-doubled Nd:YAG. .Many non-destructive testing labs use ruby lasers to create holograms of large objects such as aircraft tires to look for weaknesses in the lining. Ruby lasers were used extensively in tattoo and hair removal, but are being replaced by alexandrite lasers and Nd:YAG lasers in this application.

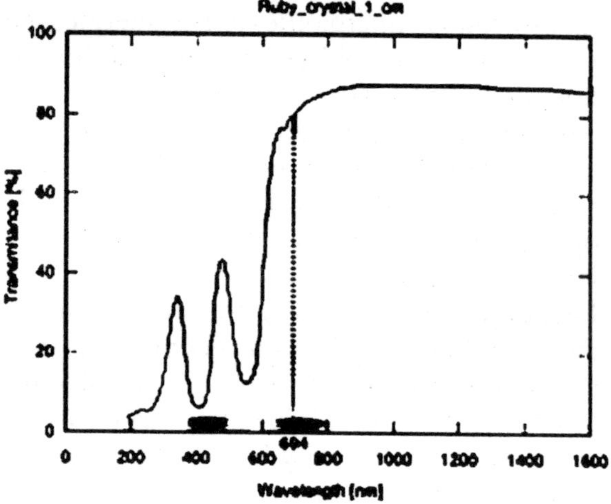

Fig. Transmittance of Ruby in Optical and Near-IR Spectra.

Design

The ruby laser is a three level solid state laser. The active laser medium (laser gain/amplification medium) is a synthetic ruby rod that is energized through optical pumping, typically by a xenon flash lamp. In early examples, the rod's ends had to be polished with great precision, such that the ends of the rod were flat to within a quarter of a wavelength of the output light, and parallel to each other within a few seconds of arc. The finely polished ends of the rod were silvered: one end completely, the other only partially.

The rod with its reflective ends then acts as a Fabry-Pérot etalon (or a Gires-Tournois etalon). Modern lasers often use rods with ends cut and polished at Brewster's angle instead. This eliminates the reflections from the ends of the rod; external dielectric mirrors then are used to form the optical cavity. Curved mirrors are typically used to relax the alignment tolerances

Coherence

Coherence is one of the most important concepts in optics and is strongly related to the ability of light to exhibit interference effects.

A light field is called *coherent* when there is a fixed phase relationship between the electric field values at different locations or at different times. *Partial coherence* means that there is some (although not perfect) correlation between phase values.

Spatial Versus Temporal Coherence

There are two very different aspects of coherence:

- *Spatial coherence* means a strong correlation (fixed phase relationship) between the electric fields at different locations across the beam profile. For example, within a cross-section of a beam from a laser with diffraction-limited beam quality, the electric fields at different positions oscillate in a totally correlated way, even if the temporal structure is complicated by a superposition of different frequency components. Spatial coherence is the essential prerequisite of the strong directionality of laser beams.
- *Temporal coherence* means a strong correlation between the electric fields at one location but different times. For example, the output of a single-frequency laser can exhibit a very high temporal coherence, as the electric field temporally evolves in a highly predictable fashion: it exhibits a clean sinusoidal oscillation over extended periods of time.

Lasers have the potential for generating beams (e.g. Gaussian beams) with very high spatial coherence, and this is perhaps the most fundamental difference between laser light and radiation from other light sources.

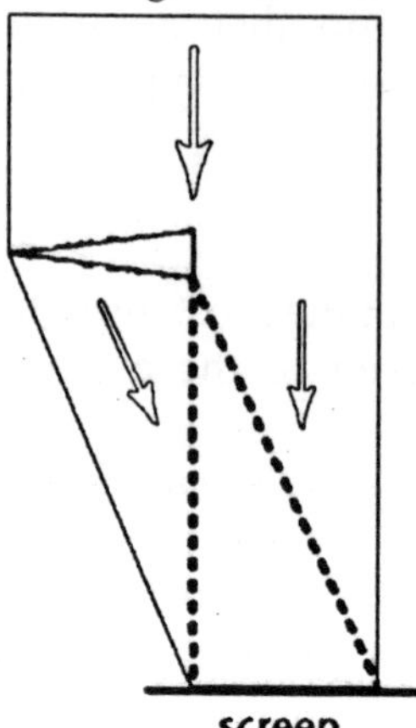

Fig. A Prism is Inserted into a Spatially Coherent Laser Beam, Generating an Interference Pattern on the Screen.

High spatial coherence arises from the existence of resonator modes, which define spatially correlated field patterns. In situations where only a single resonator mode has sufficient laser gain to oscillate, a single longitudinal mode can be selected, obtaining single-frequency operation with very high *temporal* coherence as well. The difference between spatial and temporal coherence.

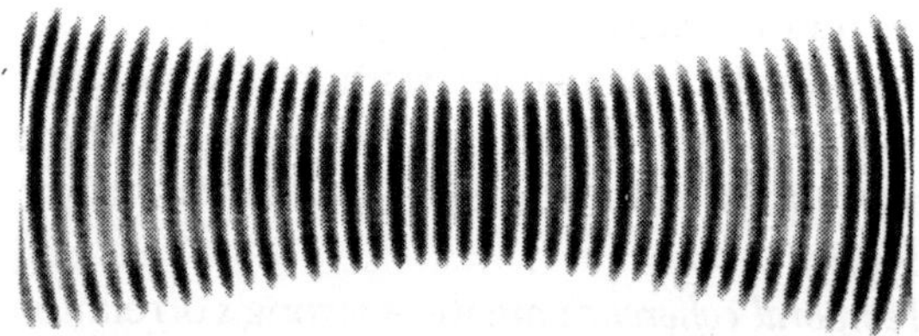

Fig. Electric Field Distribution Around the Focus of a Gaussian Laser Beam with Perfect Spatial and Temporal Coherence.

A beam with high spatial coherence, but poor temporal coherence.

The wavefronts are formed as above, and the beam quality is still very high, but the amplitude and phase of the beam varies along the propagation direction. Note that both the local amplitude and the spacing of the wavefronts vary to some extent. Such a beam can be generated e.g. from the output of a supercontinuum source.

Fig. A Laser Beam with High Spatial Coherence, but Poor Temporal Coherence.

A laser beam with poor spatial coherence, but high temporal coherence. The wavefronts are deformed, and this results in a high beam divergence and poor beam quality. On the other hand, the beam is monochromatic, so that the spacing of the deformed wavefronts remains constant. Such a beam can result from a single-frequency laser, when its output is sent through some optically inhomogeneous material.

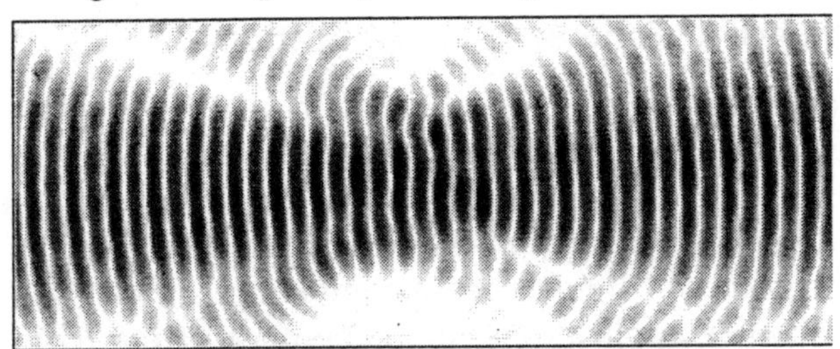

Fig. A Laser Beam with Poor Spatial Coherence, but High Temporal Coherence.

Quantifying Coherence

There are different ways to quantify the degree of coherence:

- Correlation functions specify the degree of correlation as a function of a spatial or temporal distance.
- Fringe visibility parameters essentially specify the visibility (contrast) of an interference pattern generated by superposition of two electric fields.
- The coherence time quantifies the degree of temporal coherence via the time over which coherence is lost.

- The coherence length is the coherence time times the vacuum velocity of light, and thus also characterizes the temporal (not spatial!) coherence via the propagation length (and thus propagation time) over which coherence is lost.
- The linewidth of a single-frequency laser is also strongly related to temporal coherence: a narrow linewidth (high monochromaticity) means high temporal coherence.

The relationship between optical bandwidth and temporal coherence can be non-trivial. For example, a pulse train from a mode-locked laser can have a broad overall bandwidth, with the Fourier spectrum consisting of discrete very narrow lines (→ *frequency combs*). The temporal coherence can be very high in the sense that there are strong field correlations for large time delays which are close to integer multiples of the pulse period.

Importance of Coherence in Applications

Some applications need light with very high spatial and temporal coherence. This applies, e.g., to many variations of interferometry, holography, and some types of optical sensors (e.g. fiber-optic sensors). Such features are also important for the technique of coherent beam combining.

For other applications, the coherence of the light used should be as *low* as possible.

For example, very low temporal coherence (but combined with high spatial coherence) is required for *coherence tomography*, where images are created with a kind of interferometry, and a high spatial resolution requires low temporal coherence.

Suitable light sources for such applications can be based on amplified spontaneous emission (ASE) from a laser amplifier (→ *superluminescent sources*) or on supercontinuum generation in nonlinear media. A low degree of temporal coherence can also be beneficial for laser projection displays, imaging and pointer applications, as it reduces the tendency for speckle and similar interference effects.

COHERENCE IN QUANTUM OPTICS

In quantum optics, the term *coherence* is often used for the state of light-emitting atoms or ions. In that case, coherence refers to a phase relationship between the complex amplitudes corresponding to electronic states. This is important, e.g., in the context of lasing without inversion. There is also the term "coherent states" of the light field, which has yet another meaning.

Coherence Time

The coherence time can be used for quantifying the degree of temporal coherence of light. In coherence theory, it is essentially defined as the time over which the field correlation function decays. This correlation (or coherence) function is

$$g(\tau) = \frac{\left\langle E^{x}(t)E(t+\tau)\right\rangle}{\left\langle E^{x}(t)E(t)\right\rangle}$$

where *E(t)* is the complex electric field at a certain location. This function is 1 for = 0 and usually decays monotonically for larger time delays . For an arbitrary shape of this function, the coherence time can be defined by

$$\tau_{\text{coh}} = \int_{-\infty}^{+\infty} |\, g(\tau)\,|^2 \, d\tau.$$

In the case of an exponential coherence decay (as it occurs e.g. for a laser with quantum noise influences only), this is the same as the exponential decay time.

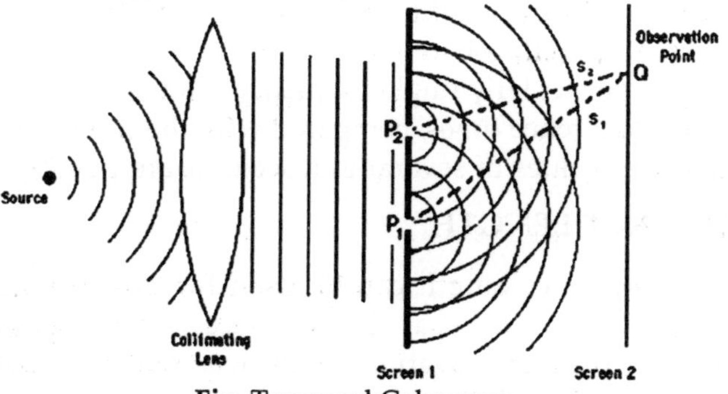

Fig. Temporal Coherence.

Knowledge of the coherence time (i.e., a single number) can be useful when the shape of the coherence function (or the shape of the Fourier spectrum) is approximately known. Obviously, however, the specification of the coherence time (or the linewidth) alone does not constitute a full characterization of the coherence.

The coherence time is intimately linked with the linewidth of the radiation, i.e., the width of its spectrum. In the case of an exponential coherence decay as above, the spectrum has a Lorentzian shape, and the (full width at half-maximum) linewidth is

$$\Delta\nu = \frac{1}{\pi\tau_{coh}}.$$

The constant factor in this equation is in general different for other shapes of the coherence function (e.g. roughly twice as high for a Gaussian shape).

Conversely, the linewidth can be used for estimating the coherence time, but the conversion depends on the spectral shape.

In cases where the frequency noise spectrum is not flat but rises strongly at small noise frequencies, there can be a significant degree of coherence even for time delays well above the inverse linewidth; this issue is important, e.g., in the context of self-heterodyne linewidth measurement.

Instead of the coherence time, it is common to specify the coherence length, which is simply the coherence time times the vacuum velocity of light, and thus also quantifies temporal (rather than spatial) coherence.

Lasers, particularly single-frequency solid-state lasers, can have long coherence times, compared with the duration of an optical cycle; values of several milliseconds are possible.

COHERENCE LENGTH

The coherence length can be used for quantifying the degree of temporal (not spatial!) coherence as the propagation length (and thus propagation time) over which coherence degrades significantly.

It is defined as the coherence time times the vacuum velocity of light.

For light with a Lorentzian optical spectrum, the coherence length can be calculated as

$$L_{\mathrm{coh}} = C\tau_{\mathrm{coh}} = \frac{C}{\pi\Delta\nu}$$

where $\Delta\nu$ is the (full width at half-maximum) linewidth (optical bandwidth).

However, such relations are not valid in cases where the coherence function has a more complicated shape, as is the case for, e.g., a frequency comb.

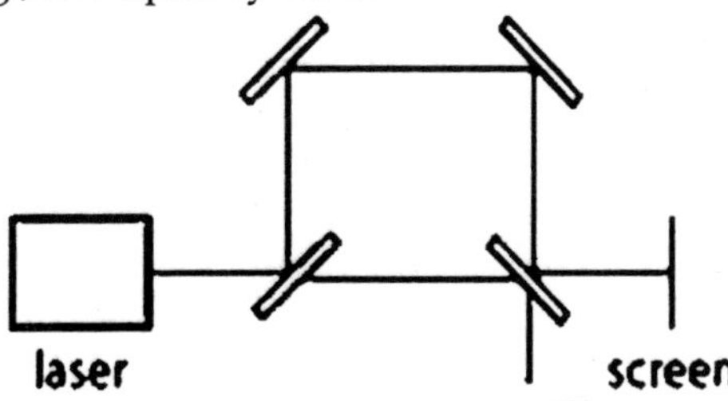

Fig. Setup of an Interferometer

The reason for often using the term *coherence length* instead of *coherence time* is that the optical time delays involved in some experiment are often determined by optical path lengths. The interferometer shows pronounced interference fringes only if the coherence length of the laser light is at least as long as the path-length difference of the two arms.

Also, in a setup for making holographic recordings, coherence between two beams with a somewhat different optical path length is required, so that the coherence length of the light source should be longer than the maximum occurring path-length difference.

In addition to holography, a number of other applications may require a certain coherence length.

Lasers, particularly single-frequency solid-state lasers, can have very long coherence lengths, e.g. 9.5km for a Lorentzian spectrum with a linewidth of 10kHz.

The coherence length is limited by phase noise which can result from, e.g., spontaneous emission in the gain medium.

COHERENCE LENGTH IN NONLINEAR OPTICS

An unfortunate use of the term *coherence length* is common in nonlinear optics: for example, in second-harmonic generation, the coherence length is often understood as the length over which fundamental and harmonic wave get out of phase (more precisely, the phase difference accumulated over this length is δ).

This is inconsistent with the general notion of coherence, because a predictable phase relationship (strong phase correlation) is definitely maintained over more than this length, although there is a *systematic* evolution of the relative phase.

Index

G

H

I

K

L

M

N

O

P